Shakespeare's Theories of Blood, Character, and Class

Studies in Shakespeare

Robert F. Willson, Jr.
General Editor

Vol. 12

PETER LANG
New York • Washington, D.C./Baltimore • Bern
Frankfurt am Main • Berlin • Brussels • Vienna • Oxford

Shakespeare's Theories of Blood, Character, and Class

A *Festschrift* in Honor of David Shelley Berkeley

EDITED BY
Peter C. Rollins
and Alan Smith

PETER LANG
New York • Washington, D.C./Baltimore • Bern
Frankfurt am Main • Berlin • Brussels • Vienna • Oxford

Library of Congress Cataloging-in-Publication Data

Shakespeare's theories of blood, character, and class: a *festschrift* in honor of
David Shelley Berkeley / Peter C. Rollins and Alan Smith, editors.
p. cm. — (Studies in Shakespeare; vol. 12)
"Publications of David S. Berkeley": p.
Includes bibliographical references and index.
1. Shakespeare, William, 1564–1616—Knowledge—Anatomy.
2. Shakespeare, William, 1564–1616—Knowledge—Psychology.
3. Characters and characteristics in literature.
4. Social classes in literature. 5. Mind and body in literature. 6. Body, Human, in
literature. 7. Eugenics in literature. 8. Blood in literature.
I. Rollins, Peter C. II. Smith, Alan. III. Berkeley, David Shelley. IV. Series.
PR3069.B58 S48 822.3'3—dc21 2001038445
ISBN 0-8204-4518-5
ISSN 1067-0823

Die Deutsche Bibliothek-CIP-Einheitsaufnahme

Shakespeare's theories of blood, character, and class: a *festschrift* in honor of
David Shelley Berkeley / ed. by: Peter C. Rollins and Alan Smith.
–New York; Washington, D.C./Baltimore; Bern;
Frankfurt am Main; Berlin; Brussels; Vienna; Oxford: Lang.
(Studies in Shakespeare; Vol. 12)
ISBN 0-8204-4518-5

The paper in this book meets the guidelines for permanence and durability
of the Committee on Production Guidelines for Book Longevity
of the Council of Library Resources.

Printed in the United States of America

Contents

Preface

Since 1972, David Shelley Berkeley has been my colleague in the English Department at Oklahoma State University—where he began his teaching career in 1948. He was a true legend, the talk of Morrill Hall: I first heard stories from faculty members and then wide-eyed anecdotes from admiring students. Later, at a graduate faculty meeting, I heard the legend speak; he corrected a full professor who mispronounced the word "enamored." Over the following months and years I frequently encountered "Shelley" in the library indeed, on many more occasions than all of my other colleagues combined. (We both had carrels at that time, a privilege jealously guarded.)

Throughout his career, Shelley has been a true "Christian Scholar." He is a believer who prays for the betterment of our world. Indeed, when Communism fell in 1989, he told me that he and his beloved wife had prayed for the demise of that Evil Empire each night for over forty years; the downfall of "the wall" was both a physical and metaphysical juncture for the Berkeleys—as, indeed, it was for the rest of us who care about freedom. In addition to being a real Christian, Shelley has been a true scion of Harvard University. He has Harvard chairs in his home and selected prints of the Harvard campus on the walls of his study. He often reminds me that he and I—as scholars—were born on "the self-same hill." The Harvard imperative to excellence has been a source of strength and an incentive to scholarly daring in his chosen areas of expertise—Milton and Shakespeare.

David Shelley Berkeley devoted his career to what we call "the life of the mind," and this collection contains essays by his devoted students. Each day—with the exception of the Sabbath—this man of faith devoted himself to the mysteries of language, literature, history, and culture. The quality of the chapters of this book speaks volumes about that dedication as well as Shelley's ability to pass along his passion to others, his true progeny. In a world of sound bites and superficial analysis, we honor a scholar not too proud to worship his Creator. He has practiced what Harvard Professor Perry Miller called "The Puritan Synthesis," a difficult task in any age—including the Puritan era. With his students, I honor that enviable balance in the life and work of David Shelley Berkeley.

Peter C. Rollins, Regents Professor of English
Oklahoma State University

David S. Berkeley

David Shelley Berkeley, Christian Scholar

I

It is difficult to write in honor of a man who disdains panegyrics.

I recall Professor Berkeley's own discomfiture when, years ago, he was asked to prepare a speech with which to honor a retiring colleague. "Panegyrics, Sir, are always to be deplored," Berkeley intoned in that grave, sonorous manner of his as we stood outside his office. "One really ought to strive to present a balanced view of any mortal." So I am bound by this remark, and bound as well by the dictates of common sense as bolstered by my own experience with human nature, to say that in David Shelley Berkeley—as in every other human being—the elements are mixed, even though my closest scrutiny of the memories I have of him fails to disclose any blot on the escutcheons either of tutelage, of friendship, or of humanity.

The unblemished condition of these escutcheons is the result, I believe, of Berkeley's resolute nature as both a Christian *and* a scholar. Being a man of faith, Berkeley is aboard the sturdy ship of a firm and abiding conviction that continues to bear him over the rough seas of life, past the wanhope and cynicism upon whose shores so many others, less safely transported, have suffered shipwreck. In his bearing, as well as in his thoughts, Berkeley is the type of what I shall ever regard as a New England Christian in the traditional sense—a man of honesty, forthrightness, and charity whose gentle sense of humor is not unlike that displayed by the Master Himself in certain passages of the Gospels.

As a scholar, Berkeley has taken up and implemented the tenets, traditions, languages, and symbols associated with Christianity as a principal means of bringing light and life to the literature of England. From his course on the literary aspects of the King James Bible, with its formidable study of KJV mistranslations of the Hebrew and Greek, to his course on the writings of C.S. Lewis, Berkeley has powerfully elucidated the many elements of Christian doctrine and tradition in which so much of British literature, from the Renaissance to the present century, is thoroughly steeped. His many publications, ranging from the early article "A Vulgar Error Touching Calvin's Doctrine of Natural Depravity" to his later book-length studies of Milton and Shakespeare, have offered practicing scholars many alternative and highly original interpretations of literary works whose themes and motifs are enriched by centuries of Christian tradition.

All of this is by no means to suggest that Berkeley has ever descended to rank proselytizing either in the classroom or in print. His spiritual temperament is broad, and his mind has always been dispassionately focused on the scholarly matters at hand. These traits, admirable in themselves, are indispensable to anyone who is charged with the task of guiding students of all faiths through the intricate connections among religion, politics, and culture that form the milieu in which so many British works of literature have been written. Such a man is David Shelley Berkeley—Christian Scholar.

II

"May you be cursed to live in interesting times," runs the Oriental malediction. Young adulthood for Professor Berkeley was undoubtedly interesting and must surely, at times, have appeared to be a curse. After graduating with an A.B. degree in German from Juniata College in his native Pennsylvania, Berkeley proceeded to Harvard University, from which he was graduated with an A.M. degree in philology (1941). The outbreak of the Second World War saw him inducted into the United States Army. It was during the induction procedure that someone employed the abbreviation *Eng* rather than *Engl* ("English") to complete a blank reserved on the induction form for listing the inductee's educational background. Berkeley would have found himself dispatched to Alaska with a contingent of engineers (Engs) had he not been stricken suddenly with influenza and confined to base, a circumstance that provided sufficient time for the military bureaucracy to discover its mistake, reassign him to an intelligence unit of the New Jersey National Guard, and send him to Alaska anyway, first to Moses Point and then to Saint Paul's Island. While stationed in these places, which saw none of the fighting that took place on the Aleutian island of Attu, Berkeley indulged in ivory carving, putting to use some of the tusks formerly sported by constituents of the indigenous walrus population, while keeping a sharp eye out for signs of a Japanese invasion that never came. A more colorful (though fictional) portrait of the young Berkeley during his Alaska tenure appears in *Guys on Ice,* a supposedly true account of the adventures of the unit to which Berkeley was attached. In this book, written by one of the unit's more self-aggrandizing personnel, Berkeley is depicted as a whiskey-drinking, battle-hardened combat soldier clenching a cigar butt between his teeth as he sprays hot lead from a Thompson submachine gun into the writhing bodies of Japanese soldiers. ("We didn't see a single Japanese the entire time we were up there," Berkeley recalled.) The library at Oklahoma State University possesses a copy of *Guys on Ice,* the title of which he has revealed only now, and I remember that he used to make regular forays into the library in order to remove that book from its proper location in the stacks so as to prevent me, or anyone else who knew of its existence, from trying to look it up.

Following his discharge from the army, Berkeley returned to Harvard and completed requirements for the Ph.D. degree with his dissertation "The Origins of Sentimental Comedy," written under the auspices of George Sherburn, the celebrated Pope scholar. Sherburn then secured for his young charge a position on the faculty of Oklahoma Agricultural and Mechanical College (presently Oklahoma State University). It was 1948, and David Shelley Berkeley—a native of Pittsburgh imbued with the ethos of the Pennsylvania Dutch and accustomed to the environs of the Eastern United States—was faced with the prospect of teaching Shakespeare and Milton at a small agricultural and mechanical college situated deep in the Land of the Red Man.

"Several of my close friends gathered round to commiserate with me," Berkeley said years later, "and I sought to prepare myself for the ordeal of geographical and cultural translation by attending a production of Rodgers and Hammerstein's *Oklahoma!* in Boston." His father, whose ideas about Oklahoma were formed during the 1890s, provided him with a pistol as a means of fending off hold-up men, lynch mobs, and hostile Indians. The elder Berkeley also refused ever to visit his son in Oklahoma on grounds that the state was both lawless and sweltering.

Assuming his duties in the English department at Oklahoma A & M, Berkeley found in his department head, Hans Andersen, a resolute opponent of any efforts on the part of faculty members to produce scholarly books and articles. Andersen sought to abort the production of scholarship by assigning Berkeley different classes every term, with each class requiring considerable time either for preparation or for the grading of papers. But Andersen had not counted on his new faculty member's German stubbornness, and Berkeley soon began to write and publish the first items in what would become a distinguished bibliography of two books and more than fifty articles. During these trying early years, Berkeley aimed his pen at any available target—including the dull and uninspiring topic of freshman composition, from whose tiresome substance at least one article, "Agreement of Subject and Verb in Anticipatory 'There' Clauses," appeared in *American Speech.* Mining the scholarly ore from his doctoral dissertation, Berkeley published three articles that are still regarded by specialists as standard works on Restoration drama: "*Préciosité* and the Restoration Comedy of Manners" in *Huntington Library Quarterly* 18 (1955), "The Art of 'Whining' Love" in *Studies in Philosophy* 52 (1955), and "The *Précieuse,* or Distressed Heroine, of Restoration Comedy" in *Oklahoma State University Arts and Sciences Studies* 56 (1959). The production of these and the majority of his subsequent articles, as well as his first book, *Inwrought with Figures Dim: A Reading of Milton's Lycidas* (The Hague: Mouton, 1974), was made yet more difficult by the debilitating illness that struck his wife, Carolin, not long after their arrival in Oklahoma. Recourse to physicians failed to produce a ready cure, and Berkeley was for years obliged to assume responsibility for her care. It was not until the late 1970s that efficacious treatment for Mrs. Berkeley's condition was at last secured.

A man of spiritual resolution, a steadfast husband, a publishing academician, and a jovial companion, all under the most trying personal circumstances—this is the Christian Scholar whom I have known since January 1973.

III

On July 1, 1987, the Christian Scholar retired from Oklahoma State University and, with his wife, began dividing his time between his Stillwater home on Lakeshore Drive (a sturdy, open-beam affair purchased in 1966 that borders on the west shore of Boomer Lake, where Berkeley still goes rowing when the Oklahoma winter permits) and his more recently acquired estate outside Red River, New Mexico. It is on this New Mexico estate, occupying three-quarters of an acre among the aspens, the Colorado spruce, and the mountains, that an essential component of the Berkeley nature comes fully into view: the Country Squire. The New Mexico manor house is a two-storied, baronial affair sporting three coats of arms downstairs and one more upstairs. The Red River itself runs through the western side of the estate. It is easy to imagine the estate's master as a Victorian figure dressed out in tweeds, Wellington boots, and a deerstalker cap who walks briskly with a silver-headed cane through the surrounding countryside. Berkeley in fact passes his Red River summers very much in this fashion, though without the Victorian paraphernalia, taking daily walks as a means of physical exercise when he is not making a broader tour of the countryside from behind the wheel of a 1973 Jeep Wagoneer. Until her passing in 1998, Carolin Berkeley devoted much of her own time during these summers to her duties as pianist at Questa Baptist Church—a position of which she was justly proud—and to preparing the Pennsylvania Dutch meals for which her husband has so pronounced a taste.

IV

The present collection of essays is, in its own way, a chronicle of Professor Berkeley's scholarly accomplishments during the mature period of his career. A bibliography that lists the publications of a practicing scholar illuminates only one aspect of that scholar's career. Another aspect, less readily visible, is embodied in the minds and souls—and therefore in the scholarly efforts—of those whom the scholar has trained. The contributors to this volume are all, in effect, the intellectual progeny of David Shelley Berkeley. In their choices of topic, in their approaches to the literary works under examination, and in the depth and breadth of their research, these figurative sons and daughters display much of their careful nurturing under the demanding yet gentle hand of the Christian Scholar. Included here is Professor Berkeley's own unpublished essay on Shakespeare and blood, which was the genesis of his second book, *Blood Will Tell in Shakespeare's Plays* (Lubbock: Texas Tech Press, 1984). The influence of Berkeley's discussion of Shakespeare's plays in the light of scientific and cultural ideas about blood current during the English Renaissance—ideas that remained prevalent well into the late nineteenth century—is evident in the majority of the contributions to this volume. But these contributions are not, I say, rank imitations; rather, they are further explorations of terrain already pioneered by Berkeley himself.

With these essays we, the contributors, hope not only to add something new to the understanding of the works of Shakespeare and other writers; we hope as well to repay in scholarly coin something of the debt each of us owes to the sound tutelage of David Shelley Berkeley, the Christian Scholar.

Alan Smith
Norman, Oklahoma
May 27, 1999

Shakespeare's *Severall Degrees in Bloud*

by David S. Berkeley

More than any English author of his time, or perhaps of all time, Shakespeare imparts the idea that "there are severall degrees in bloud"[1] to both his borrowed materials and his free inventions: he increases the distance between those of gentle blood and baseborns, and he reserves a special odium for degenerates as reflected in Sonnet 94—"Lilies that fester smell far worse than weeds."[2] His plays intensify the dominant Elizabethan view that heredity, much more than social and climatic environment, determines individuals, though genetics was subject to various exterior conditions.[3] The gentry observed the rules of eugenics as then understood, usually elaborations of Hippocrates and Galen, whereas baseborns—uninformed, impulsive, and poor—generally ignored or were compelled to ignore them.[4] Medical writers rarely discussed this subject in terms of the distinction between the gentle and ungentle classes: perhaps the verge of their thought is represented by the recommendation in *The Byrthe of Mankynde*, a technical treatise by Eucharius Roesslin, that fertility in noblewomen is promoted by "a good halfe hour" of bathing "in halfe water, and other halfe *redde wine*."[5] But writers of courtesy and conduct books were generally not slow to trace noble characteristics to gentle blood and unadmired traits to the blood of churls. Thus the author of *The Gentlemens Academie* asserts that gentry were "lordly of countenance, sweet in speech, wise in answers ... pacient in affliction"[6] and knowledgeable about their birth whereas churls were "full of slouth in warre," braggarts, cowards, lechers, drunkards, revokers of challenges, killers of prisoners, and liars to their sovereigns.

Shakespeare's plays are very much in this tradition, as Ernest Crosby[7] has noted disapprovingly without bothering with the genetic basis. Samuel A. Tannenbaum in "Shakespeare's Caste Prejudices"[8] has attempted to refute Crosby on evidence almost exclusively drawn from *Cymbeline*: he argues, for example, that Posthumus is a virtuous "son of the respectable middle class" whereas the despicable Cloten is a prince. But characters in the play whom we have no reason to distrust refer to Posthumus as a "gentleman," and only Cloten expressly regards himself as being above Posthumus by birth.[9] *Henry IV, Part I* and especially *Cymbeline* so heavily underline the belief that blood will tell that it is arguably a component of their main themes. In a non-technical way there has been some observation of the significance of heredity in Shakespeare's plays. In "Nature and Nurture in Shakespeare's Plays," Leonard Darwin takes the unexceptionable position that "Shakespeare fully realized the importance of inborn qualities"[10] though we need not follow Darwin in supposing that the poet would have been an advocate of eugenic reform. John W. Draper in "Bastardy in Shakespeare's Plays"[11] points out the Elizabethan tendency to think of the lower classes as sharing with bastards the "base effects" of congenital viciousness. G. Wilson Knight[12] has mentioned the qualities of royal blood especially to be seen in Shakespeare's last comedies, and Albert H. Tolman[13] has noticed as a property of royal blood Perdita's evidence of education without herself being educated.

In this essay I suggest that Shakespeare's plays are in alignment with the medical thought of his age; but the author, perhaps justifiably, usually suggests observance of eugenic rules in the background and behaviour of gentle characters, suspension of the effects of inadvertent malpractice among those whose blood is of excellent quality to begin with, and inobservance of the rules in the evil courses of degenerates and in hard conditions of the baseborn. Overt touches of Elizabethan eugenic thought reveal what I think are unnoticed aspects of the characterization of Orlando, Oswald, the English common soldiery in *Henry V*, Cloten, Philoten, Aguecheek, Guiderius and Arviragus, Prince Hal, Jack Cade, Falstaff, Duncan, Lady Macbeth, Hamlet, Laertes, Cleopatra, King John, Lucrece, and others.

Neglect of eugenic rules, especially those of diet, was thought to impair the fetus and to follow throughout life the individuals thus conceived. This poor start was aggravated by habitual consumption of genetically inferior foods. But if one's hereditary quality were high enough, even the consequences of base foods could be offset. Thus Orlando in *As You Like It*, albeit not the son of Sir Rowland de Boys' theoretically most vigorous blood, is a model young man even though his brother forces upon him the low-class foods of the "hinds." Oswald's lack of virtue in *King Lear* is implied by the charge that he is an eater of "broken meats." The French in *Henry V* scorn the English as drinkers of sodden barley water rather than wine, thus in effect calling attention to the superior blood of the English which enables them to prevail over the wine-drinking French nobility. Orlando is an example, to use the terminology of the author of *The Institution of a Gentleman*,[14] of "gentle-gentle" (birth and deeds); the English common soldiery (whom Henry gentles *en masse*) are examples of "ungentle-gentle" (base birth and gentle deeds); characters like Richard III are "gentle-ungentle" (gentle birth and base deeds); and, if I may coin a like expression, Jack Cade and his followers are "ungentle-ungentle" (base birth and base deeds). Ultimately the virtue of characters approved in Shakespeare's plays was the fruit of blood-inheritance. Elizabethans understood that such persons' fathers had regard to "order and consent in naturall things."[15] Their youthful parents[16] partook of goats' milk or the equivalent "six or seven daies before their companying,"[17] preferably not during the summer.[18] Copulation ideally took place at the end of the second sleep[19] and certainly not under the influence of the south wind;[20] it was violent enough to draw blood from all important quarters of the body;[21] and the prospective father should not have worn down his blood previously with excessive sex.[22] From these and other observances came thin, fast-flowing red blood and from it fine white semen of proper heat. Pregnant wives avoided sighing,[23] emotionalism,[24] phlebotomy,[25] and evil airs[26] as diminishing or corrupting the blood necessary to nourish the fetus. They avoided salt meat,[27] frightful sounds and scenes that might be reflected in the persons of their offspring,[28] and tight clothing that might deform the infants.[29] Nurses were of virtuous and benign disposition lest their milk prove corrupting.[30] Thus came first-class human beings, persons who could keep much of their innate quality even under severe deprivation. Others were not so favored. The rank smell of Cloten in *Cymbeline*[31] is presumptive evidence of his essential plebeianism despite access to the best diet. The congenital halitosis and body odor of the lower orders in *Julius*

Caesar and *Antony and Cleopatra*, on the other hand, were just what one would expect of their physiologically impoverished blood.

The purest blood in Shakespeare's characters resists the intelligence-limiting properties of the poorest foodstuffs. Cloten's customary obtuseness, which reaches a kind of apogee in his donning of Posthumus' costume (*Cymbeline* III.v.135ff), points to the causal connection between base birth and stupidity usual in Shakespeare's plays. He is told by a nameless lord, "You cannot derogate" (II.i.46), covertly implying that he cannot act in a degenerate manner because he is already as low in intrinsic quality as he can go. The intellectual grossness of the lower classes is caused in part by gross diet, such as ale, beer, apples, melons, cherries, plums, and pears. Their foods often were uncooked.[32] According to Levinius Lemnius, children who are fed "with a slender allowance and smal pittaunce of unsavoury resty flesh, stinking fish, and hoary ... bread"[33] are ugly, cowardly, and dull. Philoten's relative inability to learn, by comparison with Marina, in *Pericles Prince of Tyre*, may well be owing both to her conception by Cleon and Dionyza in famine-ridden Tarsus when the blood of her parents, although noble, had been depleted by bad diet, and to impoverishment of her own blood by the lack, when she was small, of any food except corn—unpraised by Elizabethans for eugenic qualities.[34] John Huarte pointed out the causal relationship between certain unapproved foods and beverages and the production of untowardly children:

> *Cowes flesh, manzo, bread of red graine, cheese, olives, vinegar, and water alone will breed a grosse seed, and the son engendered upon these ... be furious and of a beastlie wit .... Hence it proceedeth, that among upland people, it is a miracle to find one quick of capacitie, or towardlie for learning: they are borne dull and rude: for they are begotten of gross and evil substance.*[35]

Aguecheek's remark in *Twelfth Night*—"I am a great eater of beef and I believe that does harm to my wit" (I.iii.90–91)—reflects accepted doctrine: its humor lies in its being uttered by a congenital defective, and it would be very much to the point for Aguecheek to intimate that his parents were great eaters of beef at the time of his conception. "Hote meates," i.e. aphrodisiacs, were advised by various authorities for prospective parents because they were believed to "encrease much seed" and to draw blood for semen from all parts of the body including the remote brain. Besides goats' milk, Huarte thought that "White bread made of the finest meale, and seasoned with salt ... Partridges ... and Kid" are apt foods for parents who "shall breed children of great understanding."[36] "Fine meats," he said, are "appiable to the temperature of the wit: fore there is engendered the blood and the seed ...."[37] Here is reflected an opinion, as against Hippocrites, that semen is mostly derived from the head instead of the body generally.[38] If therefore, Huarte went on, "the meat be delicat and of good temperature, such is the blood made, and of such blood such seed, and of such seed, such braine."[39] He held that "there is no child born, who partaketh not of the qualities and temperature of that meat, which his parents fed upon a day before he was begotten."[40] Eryngoes, eggs, potatoes, pork, capon, veal, and especially wine were genetically estimable because possessing interior heat they went readily into good blood through "decoction" in the stomach, then through "sanguinification" in the liver, and then into semen by action of the organs of generation. This process did not notably detract, by reason of the heat required for decoction, from the radical and fluent heat of the body as did foods and beverages bad from a eugenic point of view. The hereditary boeotianism of Shakespeare's vulgar is evidenced in malapropisms, irrelevancy, a marked bent toward physically oriented specialties, addiction to prose, an incapacity for abstract intellection, and disregard of family. Gentles, on the other hand, could transcend deprived dietary conditions because of their eugenic inheritance; and they attested their quality by their capacity to learn as well as by their handsomeness, bravery, high regard for their families, and so forth. Orlando's diet in *As You Like It* is that of a hind, yet without schooling he is schooled. Guiderius and Arviragus in *Cymbeline* have innate knowledge, yet their food since infancy has been what the wilds provide. Prince Hal, despite much indulgence in ale-drinking, an inferior drink, possesses either angelic intuitive knowledge or innate ideas that render books and teachers nugatory. Perdita in *The Winter's Tale* has a diet identical to that of the low-born shepherdesses about her, but her easily recognized superior endowments imply high blood from birth.[41]

The purest blood in Shakespeare's characters is abundant and thin, even to old age, qualities enhanced by the consumption of wines. The lower classes in Elizabethan England could hardly ever procure wine, a deprivation whose cost they sensed. Jack Cade in *Henry VI,*

Part II shouts that when he is king of England, "I will make it felony to drink small beer" (IV.ii.73–74) and "of the city's cost, the pissing conduit run nothing but claret wine this first year of our reign" (IV.vi.3–5). Wine, as Sir Thomas Elyot intimated in *The Castell of Helth,* was almost the equivalent of blood.[42] Tobias Whitaker in *The Tree of Human Life, or the Bloud of the Grape* declared that "wines, especially Claret or red, is sanguified before it is taken."[43] Beers and ales were reckoned somewhat rebellious to assimilation into the blood and therefore required expense of irreplaceable radical as well as fluent heat for this process, if indeed they did not simply engender gross vapors.[44] Loss of heat meant a loss of blood and therefore a shortening of life. Thus the working classes, who also lost blood through sweat and tears, begot children at lower bodily temperatures and with less blood than the gentles; and the result was mediocrity of offspring or something less. Whitaker thought that "water or small-beer-drinkers, look like Apes rather than men,"[45] and their children would be the "hard-favored grooms" of Shakespeare's plays and possess "villainous low foreheads." With the blood of churls wounds would not heal easily, and there was therefore a physiological reason for poltroonery. Surely one of the most amusing aspects of Falstaff from an Elizabethan point of view, not seen (I think) in analyses of his characterization, is his substitution of sack for blood. Shakespeare's old men, including Falstaff, conventionally have little blood.[46] Unfortunately for Falstaff even sack, winiest of wines[47] and recommended for the old,[48] was an evanescent surrogate for blood; this drink gave spasmodic relief from impotence and cowardice but required constant infusions. Duncan in respect of supply of blood is Falstaff's opposite. Lady Macbeth's description of his murder—"Yet who would have thought the old man to have had so much blood in him"—rightly understood, is a great compliment, yet unobserved, to James I because his ancestor, Duncan, as an old man should have little blood and little radical heat. His miracle of copious free-flowing blood in old age implies the virtue of Duncan's ancestry and his own virtuous life. It implies that Duncan could beget children, males like himself even in his old age, and thus transmit to his posterity qualities for which he serves the throne of Scotland. Duncan's effusion of blood is a sign that Malcolm will be much like his father. It is quite wrong to dismiss this line, as A.C. Bradley does, as merely part of the gory atmosphere of *Macbeth.*[49] Duncan's blood characterizes Lady Macbeth as a much younger person whose blood is ironically so thick with melancholy and so diminished in quantity that when she utters this speech she is tottering in mind and body. Indeed, there is a suggestion of the effects of diabolism in the sleep-walking scene. Devils sucked blood, as Joan of Arc told the assembled fiends (*1 Hen. VI*.V.iii.14); Lady Macbeth's disordered mind, as well as her childlessness, probably originates in part from the diabolic removal of much of her blood, which by its diminution could scarcely support a fetus. Shakespeare, of course, neither in *Macbeth* nor even in *1 Henry VI* goes to the length of Henry Goodcole in *The Wonderful Discoverie of Elizabeth Sawyer a Witch, Late of Edmonton*[50] in describing precisely how devils (or demons) remove blood from human females. The posterity for which Macbeth so inordinately hopes has, thus considered, an unnoticed connection with the witches and their familiars. One must observe that Shakespeare scatters suggestions of demon-possession over high and low in the social structure.

The purest blood in Shakespeare's characters is the most resistant to toxic effects. Hamlet has been wounded with a rapier whose venom is more virulent because it is fresh; yet he possesses the energy to pierce Laertes with the weapon whose poison is diluted with Hamlet's own blood, and Laertes dies first. Hamlet, though for some time envenomed and fatigued, is yet able to medicine his stepfather like a dog. Since Shakespeare is in these details not following a known source, the times and circumstances of the two deaths imply (among other things) that the poet apprehends "severall degrees in blood." Hamlet assuredly does not die last to illustrate the point that his blood is noblest. He remains to do what he has to do. But if his blood is equal to or less than Laertes' blood in resistance to toxicity, he should have died first. The fact that Hamlet dies last underscores the point that his blood is less like poison than that of Laertes on this account (as well as in other matters)—an observation that I believe has not emerged in criticism of the play. Shakespeare's plays offer similar illustrations of degrees in blood. In the last scene of *Antony and Cleopatra* the queen's high blood is intimated by her requiring the biting of two asps to expire whereas one asp's biting kills Iras, and another's kills Charmian—three details not found in Plutarch. In the fifth act of *King John* the King and a monk, his taster and poisoner, both drink of the same poison; but the effects, which are Shakespeare's and not Holinshed's, differ notably. Hubert says

(V.vi.31–32): "A monk, I tell you, a resolved villain, / Whose bowels suddenly burst out. The King / Yet speaks and peradventure may recover." C. E. Moberly notes, "It is strange that Shakespeare should make the monk's 'bowels burst out' when nothing of the kind happened to John."[51] The point is that John, actually a somewhat degenerate king as the fifth son of an obviously masterful, evil woman, still has a much higher genetic inheritance than a plebeian monk; so his blood, being very unlike poison, resists poison more vigorously.[52]

Degeneracy of the individual's postnatal life might come from a variety of living conditions. Andrew Boorde in *A Dyetary of Helth* pointed out that "evyl and corrupt ayres doth infecte the blode,"[53] breeding many diseases that shorten men's lives. "Standing waters, stinking mists, or marshes" were places to avoid in building one's house.[54] Moderate sleep, said Boorde, nourishes the blood; but immoderate sleep lightens the brains, engenders "imposthumes," and instigates sin.[55] Degeneracy could derive from "gluttony, or immoderacy in sucking, drinking, eating, moving, or sleeping" according to John Makliure;[56] and intemperance in sex, as Thomas Cogan[57] noted, was harmful. Antony's loss of acumen is a case in point. The harboring of envy, hatred, persecuting spirit, and pride by reducing the amount and quality of blood led to degeneracy. The sighing and weeping brought on by illicit sexual relations consumed some blood and corrupted the remainder. For example, in *The Rape of Lucrece*, the knife being withdrawn from Lucrece's breast, two streams of blood issue therefrom—one still red and pure and the other black and coagulated, having been stained by Tarquin.[58] Wherever in Shakespeare works a person of high degree acts in an unbecoming manner, one may on Elizabethan assumptions suspect bastardy in which the mother's base blood has prevailed at conception over the noble father's, or there may have been an exchange of infants, or degeneracy—prenatal and/or postnatal—has set in. There may be other Elizabethan explanations such bad stars, but this one certainly needs considering. Elizabethans would have wanted some accounting for anomalies like Claudius, Goneril and Regan, Richard III, Richard II, Oliver and Duke Frederick, Proteus in *The Noble Kinsmen*, and Antonio in *The Tempest*.

Plebeians, for various reasons, some already indicated, had almost insuperable difficulties in enhancing the quality of their blood. In Shakespeare's plays they may be loose in sexual arrangements, ignorant of the principles of eugenics, and creatures of an enforced genetically inestimable diet. They are "weeds," as Shakespeare says; but the system is not quite immutable. Helena, for example, in Shakespeare's revision of Boccaccio is so rich in "virtue dative" that the King of France voluntarily offers to raise her from the ranks of the non-titled gentry to the nobility. One is obliged to point out that the nameless First Servant of *King Lear* fatally wounds his master, the Duke of Cornwall, for blinding Gloucester: Elizabethans might have accounted for his unparalleled courage and honor by supposing that he was a bastard from a father like Richard the Lion-Hearted; it was not like them silently to accept as unexplainable practical deviations from theory.

Shakespeare is in the tradition of the "rich blood of the Round Table," making much, as Malory did, of the distinctions between nobles and churls. The motif of nobility in the peasant way of life, seen in accounts of Beaumains, Tor, and Brunor in *Morte Darthur* and in Horn in *King Horn*, is reflected in various ways in Prince Hal, Perdita, Guiderius and Arviragus, Posthumus, and Orlando. Shakespeare is quite different from Marlowe, Dekker, and Webster, all of whom emphasize the ability of baseborns to rise rapidly to positions of highest eminence. He is different from Chaucer in that the elder poet thought of "gentillese" as something that commoners might share with the bloods. Shakespeare's well-known propensity for mixing dramatic kinds tends to obviate the idea that his blood-consciousness is really a deferring to a generic stereotyping of royals, nobles, other gentles, and plebeians. It is, I believe, correct to say that all his finished products—nearly forty plays—exhibit more social distance between gentry and plebeians than do the sources from which he borrowed them. Even in his Roman plays, where he most closely adheres to ancient and therefore highly respectable sources, there is deliberate heightening of these distinctions as in, for example, the poet's introducing the peasant with the asps into *Antony and Cleopatra.* If favoring of the gently born was conventional, as some have thought, Shakespeare was conventional to the point of idiosyncracy. The egalitarian masses of the modern world and many of their scientific preceptors dislike his aristocratic notions when they are able to perceive them and believe that experience and scientific truth run counter to them; but some scientists assert that they have evidence that heredity more than social and environmental factors determines individuals.[59] Whatever is thought of the physiology implicit and at times explicit in Shakespeare's plays, whether or not the poet's eugenic bias rests upon truth is yet unknown.

Notes

1 Francis Markham, *The Booke of Honour: or Five Decads of Epistles of Honour* (London, 1625) 46.

2 All Shakespearean quotations follow *Shakespeare the Complete Works*, ed. G.B. Harrison (New York: Harcourt, Brace and World, 1952), in text and lineation.

3 Cf., for example, Markham 38. A minority view, expressed by Thomas Miles in *The Catalogue of Honor* (London, 1610) sig. B 2, held that virtue "dative," referring to the first generation to be "gentled" or ennobled, is far more excellent than virtue "native."

4 Alexander Read's preface to Owen Wood's *An Alphabetical Book of Physical Secrets* (London, 1639) suggests the rarity of authoritative books of medical self-help available to those who had neither physician nor apothecary.

5 (London, 1560) sig. xxxv.

6 (London, 1595) sig. Niii^{r}–Niii^{v}

7 "Shakespeare's Attitude toward the Working Classes" in *Tolstoy on Shakespeare* (New York: Funk and Wagnalls, 1906) 127–65.

8 *Shakespearian Scraps and Other Elizabethan Fragments* (New York: Columbia University Press, 1933) 132–76.

9 David Berkeley and Donald Eidson, "The Theme of *Henry IV, Part I*," *SQ*, 19 (1968): 25–31.

10 *Eugenics Review* 19 (1927): 189.

11 *Shakespeare-Jahrbuch* 74 (1938): 189.

12 *The Crown of Life* (London: Methuen, 1948) 162.

13 "Is Shakespeare Aristocratic?" *PMLA* 39 (1949): 288.

14 (London, 1555) sig. B 6^{V} ff.

15 John Huarte, *The Examination of Men's Wits*, trans. Richard Carew (London, 1594) 20.

16 A widespread explanation of degeneracy was the overaged father. Dr. John Makliure, *The Buckler of Bodilie Health* (Edinburg, 1630) 84, held that the prime of man's life for purposes of reproduction was between the ages of twenty-five and thirty-five. He wrote: "… bairnes procreate of bairnes, or old men, are commonly infirme, either of body or mind, but begotten in the flower of the age when the body and the spirit are at the

best, are found to be most able for any business." Cf. Thomas Cogan, *The Haven of Health* (London, 1589) 68.

17 Huarte 305.

18 Cf. William Vaughan, *Approved Directions for Health* (London, 1600) 134; Cogan 256.

19 Makliure 72.

20 Andrew Boorde, *A Compendyous Regyment or a Dyetary of Helth* (1542) EETS Extra Series No. 10 (London: Kegan Paul, Trench, Trübner, 1870) 238.

21 Levinius Lemnius, *Touchstone of Complexions,* trans. T. Newton (London, 1565) 2.

22 Makliure 72.

23 Cf. "blood-sucking sighs" in *3 Hen. VI,* IV.iv.22.

24 Makliure 124.

25 Nicholas Gyer, *The English Phlebotomy* (London, 1592) 104.

26 Jacques Guillemeau, *Chyldebirth, or the Happy Deliverie of Women* (London, 1602) 18–19.

27 Guillemeau 20.

28 Guillemeau 26. See also *The Hermetic and Alchemical Writings of Aureolus Philippus Theophrastus Bombast, of Hohenheim, Called Paracelsus the Great,* ed. Arthur Waite (London, 1894) 121–22, 315.

29 Guillemeau 26.

30 Jacques Guillemeau, *The Nursing of Children* (London, 1612) 4; Thomas Phaire, *The Boke of Chyldren* (London, 1543) 18; John White, *English Paradise Discovered* (London, 1612) 15.

31 Cf. II.i.17.

32 Cf. C.H. Simpkinson, *Life and Times of William Laud Archbishop of Canterbury* (London: Murray, 1894) 23–24.

33 Lemnius 44.

34 *Pericles Prince of Tyre,* I.iv, indicates that Cleon and his wife, together with the people of their country, have been starving to death for two years. This situation was relieved to a degree by Pericles' shipload of corn. But Elizabethans nowhere recommended cereals as an aid to procreation or as a notable stimulant to health. Indeed, Elyot (37) writes of barley corn as cooling the body, in effect reducing sexual desire. In this stressful time Philoten was conceived because she is about the same age as Marina, who was conceived shortly after Pericles' shipwreck and rescue at Pentapolis. See Walter L. Baker, "The Physiological Basis of Learning

Ability in Shakespeare's *Pericles Prince of Tyre,*" M.A. thesis, Oklahoma State University, 1971.

[35] Huarte 305.

[36] Huarte 303–4.

[37] Huarte 303.

[38] Huarte, Cf. R.B. Onians, *The Origins of European Thought* (Cambridge: Cambridge University Press, 1954) 121.

[39] Huarte 303.

[40] Huarte 295. Cf. Cogan 241; Roesslin sig. xviii.

[41] William Younger, *Gods, Men, and Wine* (Cleveland: World Publishing Co., 1966) 486.

[42] (London, 1541) 47–48.

[43] (London, 1638) 30.

[44] Whitaker 36.

[45] Whitaker 31.

[46] Cf. *The Winter's Tale,* II.ii.165–67; *Troilus and Cressida,* I.iii.291–301; *Titus Andronicus,* III.i.66.

[47] Cf. John Taylor, *Drink and Welcome* (London, 1637) sig. $B3^v$–$B4^r$.

[48] Vaughan 11.

[49] *Shakespearean Tragedy* (New York, 1966) 281. On this point I acknowledge indebtedness to Wanda M. Thilsted, "Infinite Generations," M.A. thesis, Oklahoma State University, 1970.

[50] (London, 1621).

[51] C.E. Moberly, *Rugby Shakespeare* (1883), as quoted by H.H. Furness, *A Variorum Edition of Shakespeare's* The Life and Death of King John (Philadelphia and London: Lippincott, 1919) 404.

[52] I express indebtedness for this insight to my former student Miss Russell Lang.

[53] Boorde 235.

[54] Boorde 238.

[55] Boorde 235–36.

[56] Makliure 127.

[57] Cogan 256.

[58] Ll. 1774–1750.

[59] Cf. "History and the Genes," *Time* 17 Oct. 1969.

Of Lively Grapes and Windy Hops: Blood and Drink in Renaissance English Literature

by Alan Smith

Renaissance images of blood are images of drink. The dramatists and poets of the era tended almost invariably to describe blood in terms of either malt or, especially, vinous liquors. So strongly identified is blood with drink that in many images all distinction between them disappears. For instance, Stuart drama—always noted (and for years critically deplored) as a repository of gore—abounds in images of blood and bloodshed so forcefully enhanced by, or so deeply couched in, allusions to wine that the wine in fact becomes blood. In Part I of Dekker's *The Honest Whore,* Matheo fears that Hippolito will "bleed three pottles of Alicant" in swordplay with the Duke and his cronies (I.i.86), and later Fustigo refers to the blood flowing from his own scalp by saying, "This should be a coronation day; for my head runs claret lustily" (III.i.202–3). In Shakespeare's *Macbeth,* Duncan's blood becomes a symbol of universal emptiness when Macbeth exclaims, "The wine of life is drawn, and the mere lees / Is left this vault to brag of" (II.iii.100–1). In Marston's *Antonio's Revenge,* Piero draws himself a flagon of bordeaux with the grim jest, "'Tis well, brim full. Even I have glut of blood" (V.iii.23), and in Ford's *The Broken Heart,* Orgilus' blood "sparkles like a lusty wine new broacht" (V.ii.126).

These and other images equating blood with wine convey to modern sensibilities perhaps only the figurative technique of metaphor.

Yet to the authors and their contemporaries, the equation was literal: wine, a staple of the Renaissance pharmacopoeia, was regarded by the medical community as an enhancer of, and even a surrogate for, blood.

The equation of blood with wine was scientific in origin, deriving from Hippocratic and Galenic tradition and infiltrating the popular imagination by way of the scores of vernacular encyclopedias available to the literate segment of the public. A look at the widely disseminated ideas in these texts indicates that men and women who could afford to do so drank wine regularly as a means of replenishing and fortifying their blood so as to endow themselves with amiable dispositions, courage and fortitude, fertility, keen mental acuity, and long life. The sanctions these texts imposed on the habitual consumption of beer, ale, perry, and other non-vinous beverages suggest that men and women who frequently indulged in them were depleting and corrupting their blood, rendering themselves cowardly, boorish, disagreeable, dull, and short-lived. The dramatists and poets of the time quite naturally found in this rich body of medical lore a technical rationale for the creation of images and the depiction of characters.

Renaissance dietetics—such as we may say the science existed at the time—was based upon the principle that diet, of which alcoholic beverages constituted a significant part, determined the quantity and quality of blood contained within the body.[1] From chyle (food and drink broken down in the stomach), the liver prepared veinous blood consisting of four humors—blood, choler, melancholy, and phlegm—that preserved and nourished the body's various organs and members. Each humor was either hot or cold and moist or dry, depending upon the nature of the particular organ or member for which it was ordained. Timothy Bright in *A Treatise of Melancholie* provides a typical description of veinous blood:

> *The purest part which we call in comparison and in respect of the rest bloud, is temperate in qualitie, and moderate in substance, exceeding all the other parts in quantitie, if the bodie be of equall temper, made for nourishment of the most temperate parts, and ingendering of spirits. The second is fleume, next to bloud in quantitie, of a waterie nature, cold and moyst, apt to be conuerted into the substance of purebloud if nature faile not in her workinge, ordained for nourishment of moyster partes. The thirde is melancholie, of substance grosse and earthie, cold and drie in regard of the other, in quantitie inferiour to fleume, fit nourishment for such partes as are of like temper.*

The fourth, choler, fierie, hote, and driest of qualitie, thinne in substance, least in quantitie, and ordained for such parts as require subtiller nourishment, and are tempered with greater portion of the fierie element.[2]

To paraphrase Robert Burton in *The Anatomy of Melancholy*: blood is the warm, moist humor from which the heart refines vital spirits; phlegm, cold and moist, nourishes the cold, moist parts of the body, such as the muscles and the tongue; choler, which is hot and dry, preserves heat in the bowels; and melancholy, the coldest and driest of the humors, nourishes the bones.[3]

As Bright's analysis of the blood suggests, there was a definite balance of humors thought to exist under ideal conditions: blood formed the major portion, followed by choler, melancholy, and phlegm. Unfortunately, conditions were seldom ideal, and such factors as heredity or the influence of the stars at birth could upset the ideal balance so that one of the baser humors became predominant. The humoral predominance determined the personality, or "complexion," of the individual human being so that his or her psychological profile reflected the characteristics of the predominating humor. Sanguine people, those fortunate few in whom warm, moist blood was predominant, were inclined toward conviviality, sensuality, close friendship, and courage; cholerics, whose blood was composed primarily of hot, dry choler, were prone to storms of wrath and anger; melancholics, whose blood was cold and dry, were inclined toward sadness, envy, grief, and brooding; and phlegmatics, whose blood was cold and moist, were given to laziness and daydreaming.

One way of regulating the balance of humors and so exercising some degree of control over complexion was through attention to diet. Good chyle begot much humoral blood, and much humoral blood meant a sanguine complexion. People took care to avoid foods and drinks that engendered veinous blood containing a predominance of one of the baser humors, especially melancholy, which Thomas Walkington describes as "the most vnfortunate and greatest enemy to life, because his qualities being cold and drie do most of al disagree from the lively qualities, heat and moisture: either with his coldnes extinguishing naturall inherent heate, or with his drines sucking vp the natiue moisture ...."[4] Elizabethans and Jacobeans presumably limited their intake of old beef, which Sir Thomas Elyot says "maketh grosse bloudde, and ingendreth melancholy,"[5] and they presumably followed Timothy Bright's advice largely to forego "porke, except it be yong, and a little corned with salt, beefe, ramme mutton, goate, bores flesh & veneson: neither is mutton of anie sort greatly commended of Galen."[6]

Among alcoholic beverages generally proscribed because of their cold, dry, melancholic properties were beer, ale, cider, and perry. James Hart points out in KLINIKE, *or The Diet of the Diseased* that "Thicke and muddy drinke is altogether unwholesome, and the mother of many diseases: and that which is cleare, is best and wholesomest; providing alwaies, that water be not too much master,"[7] a principle Robert Burton follows in condemning thick, dark beers as being "too windy, because of the hop."[8] The wind engendered by thick beers, evidenced by belching and farting on the part of the drinker, cooled and dried the blood, depriving it of heat and moisture. Hart claims that "Thicke beere ingendreth wind, all manner of obstructions, the stone, strangury, and many more dangerous diseases,"[9] although he admits that the windy properties of beer may in certain cases prove ameliorative: the wind inspired by the hop "openeth the obstructions of the liver, spleen, and kidnies, cleereth the blood, and cleanseth choler .... But if there bee too much hop in [the beer], or yet [the] drinke [be] very new, the hop will wrong the head, by sending up to it hot exhalations, and so procureth rheumes; and the bitterer the Beere is, the lesse it nourisheth."[10] Ale was regarded as being "of a grosser substance [than beer], and nourisheth farre more; but is of a more oppilative and stopping faculty: and therefore such as are obnoxious to obstructions ought to bee sparing in the use of this drinke."[11] Cider and perry, Burton says, "are both cold and windy drinks,"[12] but Hart recommends them "for hot and dry cholerick bodies, hot livers, and melancholicke persons: they are of an opening and penetrative nature, opening obstructions notably," provided they are aged at least four to five months.[13]

Unlike beers and ales, wines were believed to replenish and heat the supply of humoral blood and therefore to enhance its suitability for the production of vital spirits, the source of human heat and energy. Following its preparation in the liver, veinous blood separated into its constituent humors, each of which traveled through the veins to particular organs. Choler traveled to the gall bladder; melancholy went to the spleen; phlegm retired to the kidneys. Blood, the most sovereign of the humors, proceeded by way of the vena cava into the right ventricle of the heart, where it was stored and slowly transferred through a septum into the left ventricle. Here it was refined into vital spirit. Robert Burton describes the left ventricle of the heart as "the seat of life, which, as a torch doth oil, draws blood unto it, begetting of it spirits and fire; and as fire in a torch, so are spirits in the blood ...."[14] This subtler heat and moisture—that is, the vital spirits—engendered from humoral blood are the "two pillers" of life to which Andre du Laurens refers in *A Discovrse of the Preseruation of the Sight*:

> *Our life is stayed vpon two pillers, which are the radicall heate and moisture; the radicall heate is the principall instrument ... that concocteth and distributeth our nourishment, which procureth generation, which stretcheth out and pearceth the passages, which fashioneth all our parts, which maketh to liue (as saith* Trismegistus*) all the seuerall kindes of things that are in the whole world, and gouerneth them according to their worth and dignitie. This heate being a naturall bodie hath neede of nourishment, the humour which is called the radicall moisture, is the nourishment thereof, as the oyle which is put into the lampes, doth maintayne and feede the flame ....*[15]

Hence, "Blood is the oile of the lampe of our life," as Thomas Walkington says, for it is the source of vital spirits that carry heat and energy through the arteries to all parts of the body.[16]

The heat and moisture imparted by wines produced humoral blood highly suitable for the production of vital spirits. "The best remedy to recouer vitall spirites decaied," claims Nicholas Gyer in *The English Phlebotomy*, "is drinking of wine: for wine among things nourishing quickly and in short time is the most principall."[17] Wine "purges out the feculent lees of melancholy, refines and purifies the inward partes, opens the obstructions of the veines, like *Medeas* drugs, makes one young againe," according to Walkington.[18] Drinkers were supposed to be careful to select wines that had been properly aged, for, as William Turner notes in *A New Boke of the Natures and Properties of All Wines*, "The Wines that are to olde or to newe, are to be eschued" because "the olde doe heate to much, and the newe Wines as long as they are greene, or very new, heat nothing at all ...."[19] James Hart points out that "The yellow wines are the hottest, the red lesse; and the white least of all," and that thick, dark wines "are of all others aptest to nourish and fatten the body," although overindulgence in them may promote "obstructions, the stone, the dropsie, &c."[20]

The kinds of drinks indulged in depended upon the individual complexion—the amount of heat and moisture present in the body to begin with. Sanguines and especially cholerics, whose blood was already hot, naturally drank wines in moderation because to overindulge in them would be to generate excessive heat within the body and so to risk premature death as a result of damage to the organs or the exhaustion of radical heat and moisture. Thus Timon's curse on the *banditti* in Shakespeare's *Timon of Athens*: "Go suck the subtle blood o' the grape / Till the high fever seethe your blood to froth, / And so 'scape hanging" (IV.iii.432–34). On the other hand, phlegmatics and melancholics, whose blood was cold and dry, were encouraged to drink wines so as to enrich their blood with heat and moisture. The melancholic protagonist in Jonson's *Volpone*, having experienced fear that has cooled his blood even further, calls for "a bowl of lusty wine, to fright / This humour from my heart," observing after a few drinks that "This heat is life; 'tis blood by this time!" (V.i.11–12, 17).

Since the complexion was thought to change with age, declining from the heat and moisture of youth to the cold, dry, melancholic state of old age, it was axiomatic during the Renaissance that the young should drink ales and beers and that the old should drink wines. Falstaff's remark in Shakespeare's *Henry IV, Part II* that young Prince John "drinks no wine" and "There's never one of these demure boys come to any proof, for thin drink doth so overcool their blood" (IV.iii.96–98) is squarely based upon contemporary medical thought. Thomas Thompson in *A Diet for a Drvnkard* observes that "all Philosophie and Phisicke denieth wine, or strong drinke (especially in excesse) vnto young men, because it will by inflamation drie vp and consume their naturall heate, as fire put to fire,"[21] and James Hart points out that even "strong Ale and Beere … are not good for young people, hot and cholericke complexions, and hot seasons of the yeere" because they generate inordinate heat that exhausts the supply of blood and spirits.[22] Robert Burton especially disallows "All black wines, over-hot, compound, thick drinks, as muscadine, malmsey, alicant, rumney, brown bastard, metheglin, and the like" for "such as are hot, or of a sanguine, choleric complexion, young, or inclined to head-melancholy."[23] Old people, on the contrary, were encouraged to drink these and other "hot" beverages because they heated and enriched the blood, providing the warmth and moisture necessary for the production of the vital spirits that sustained life.

"Muscadines, sweet Canary sackes, and sweet nourishing red wines" were especially recommended for the old.[24] Wines were prescribed for the elderly even at breakfast: James Hart says that although drinking in the morning—regarded by the ignorant as a practice that "washeth down all ill humors, that lie on the stomack"—is in fact "very hurtfull to the health,"[25] old people should breakfast upon "wines, muscadine, sweet canary, and the like, with a tost soaked in them and eaten."[26]

For those who could afford such restoratives of blood there was *aurum potabile,* or drinkable gold, a distillation of gold usually taken with wine or some other alcoholic beverage. A typical formula for *aurum potabile* is Roger Bacon's:

> *Take wedges of fyne golde, well warmed in the fyre, and quenche it xlii tymes within the water of a good welle or fountayne. And the water strayned and kepte clenly in a vyoll of glasse, & let the dyseased person drynke it with wyne or good ale after as the pacient it requyreth, the vertue & propertie of that water of golde is suche that it is good vnto ye people that be hoote, and also for them that be sycke of the syckenes, that men called, cardia passio, for it exalteth & comforteth the herte merueylously, it illumineth the spirites, and consumeth ye superfluyte, and comforteth the vertue conteyned in all the membres within forth, & moche it auayleth vnto the fluxe of the wombe, more than the water wherein yron is put, it auaileth moch vnto the leprous, & vnto them the whiche ben yll dysposed.*[27]

A later formula developed by Francis Anthonie had a vogue during the early seventeenth century and in fact remained popular well into the Restoration.[28] Anthonie's formula was compounded of "physicall Tincture of gold," prepared as follows:

> *First, when the gold is dissolued in the proper menstruall or instrumentall water, this water is distilled away from the dissolued gold, vntill it be in the bottome in the forme or consistence of a liquid gum or hony, deepely coloured ... & most vnctuous. To which, the spirit of wine is then put, and by solution thereof, tincted very red and deepe coloured. That tincted spirit of wine is gently decanted, or powred off, and so reserued by it selfe in a glasse.*[29]

One ounce of this tincture was added to sixteen ounces "of generous wine (or if you please, of any other liquor)" and taken in doses of one to two spoonsful in an ounce of wine.[30] Anthonie claims that gold "being made potable ... restoreth, augmenteth and preserueth that our *humidum radicale,* and *calidum innatum,* Naturall heate and moysture ... and so with respectiue possibilitie, prolongeth life, and keepeth backe old age, I meane the sense of the defects and infirmities thereof."[31]

We might speculate, of course, that the rejuvenative effects of *aurum potabile* were more directly attributable to the "generous wine" or other beverage with which it was taken. This was almost certainly true of John Woodall's *aurum vitae,* a restorative that "performeth its operation only by sweat"[32] and that was sold in the form of pills advertised as safe enough for infants ("it is small in quantity, and is without offence in taste or smell to any") and pregnant women.[33] Touted as a remedy for smallpox and bubonic plague, *aurum vitae* was best taken "either mixed with some cordiall thing ... or ... with the pap of an Apple," after which the patient should take "a reasonable full draught" from "an ordinary posset of Ale and milke to the quantity of a quart very warm,"[34] then lie covered in blankets, "and so in Gods name let him lye still and sweat gently" for three to four hours, refreshing himself with the ale and milk as thirst required.[35]

The production of good humoral blood by such measures as these assured not only the production of good vital spirits, which endowed the body with strength and vitality, but also the production of good animal spirits. Once engendered in the heart, vital spirits swept through the arteries to all parts of the body, including the brain, which according to Robert Burton "hath many concavities distinguished by certain ventricles, which are the receptacles of the spirits, brought hither by the arteries from the heart, and are there refined to a more heavenly nature, to perform the actions of the soul."[36] This "heavenly" substance, animal spirit, was the mediator of perception, thought, moral judgment, and action. As such, it provided Renaissance theorists with an equivalent of the nervous system as well as with a metaphysical link between the mortal body and the immortal soul. Moralists as well as physicians therefore dealt extensively in their writings with the effects of diet, especially as it included alcoholic beverages, on human health and behavior.

Physiologically, the ingestion of alcohol contributed to the amount of chyle which the liver had to refine into humors. Overindulgence obviously resulted in the accumulation in the liver of a "great abundance of rawe & vnconcocted humors"[37] that choked the veins and impeded the distribution of humoral blood, bringing on what J. Oswen in *An Inuectiue agaynst Dronkennes* calls "ye iuste punyshment of dronkenes, great hed ache and debilitation of all ye members"—in other words, a hangover. "On ye morrow," Oswen continues, "ache and wine not as yet digested, kepeth [the inebriate] in bed tyl none."[38] According to James Hart, chronic heavy drinkers suffer from such ailments and conditions as apoplexy, epilepsy, "*Incubus* or *nightmare*," palseys, giddiness, lethargy, and loss of memory. These afflictions, caused by a plethora of unrefined or polluted humoral blood, are writ large on the drunkard's appearance in the form of "red and watery eyes, a corny face, all beset with rubies and carbuncles, accompanied with a copper nose."[39]

Spiritually, the physiological effects of excessive drinking interfered with the various operations of the rational faculty, especially the understanding and the will—the two powers most involved in rational and moral behavior. Polluted humoral blood begot polluted vital spirits which, in their turn, begot polluted animal spirits—and animal spirit was the fuel by means of which the rational faculty operated. Fueled by animal spirits ultimately refined from impure blood or from blood reeking with the fumes of alcohol, the understanding and the will could not adequately counter the effects of passions. The inebriate was thus left at the mercy of inordinate desires and fears.

In and of themselves, passions—as the moralist Thomas Wright observes—"trouble wonderfully the soule, corrupting the judgment & seducing the will, inducing (for the most part) to vice, and commonly withdrawing from vertue"[40] by reason of the nearly overwhelming arousal of humors that accompanied them. For instance, "hot" passions such as anger incited the gall bladder to release a plethora of choler, whose heat could so inflame the blood and spirits as to scorch the brain and render the understanding and the will inoperable. With the instruments of rational and moral behavior thus impaired, the individual would act according to the nature of his or her feelings, perhaps transgressing civil as well as moral law by committing anything from adultery to a murder "in the heat of passion." Similarly, "cold" passions such as envy prompted the spleen to release melancholy, whose chilling influence could freeze the blood and spirits, thereby refrigerating the intellective and moral faculties. The individual's behavior, in such a case, might be despair-induced suicide or else cold-blooded murder. Consequently, Renaissance moralists and physicians often conceived of passions as "maladies, or sores of the soule"[41] because of the sinful nature of actions committed under their influence. The souls of inebriates, who deliberately surrendered themselves to passions by numbing their rational and moral faculties with drink, were accordingly thought to be greatly diseased. Drunkards therefore were condemned by J. Oswen as "an vnprofitable burden of ye earth,"[42] by James Hart as habitual breakers of all Ten Commandments,[43] and by Thomas Thompson as beasts who "cannot possibly bring forth any other fruit but blasted, drie and withered ...."[44]

The various physical, mental, and moral aberrations of drunkards were transmitted to their offspring via blood in the form of seed. According to Renaissance medical thought, both parents, during intercourse, contributed seeds that united and decomposed within the mother's womb so as to form an embryo. Within these seeds were the *rationes seminales*, or seminal reasons, in effect a genetic program directing the seeds to decompose into an embryo that reproduced the physical and psychological traits of both parents. As with every other fluid in the body, male and female seed were thought to be refined from blood. Thomas Cogan in *The Haven of Health* says that after the third and last concoction

> *which is doone in euerie part of the bodie that is nourished, there is left some part of profitable bloud, not needefull to the partes, ordeyned by nature for procreation, which by certaine vesseles ... seruing for that purpose, is woonderfullie conueighed and carried to the genitories, where by their proper nature that which before was plaine bloud, is now transformed and changed into seede.*[45]

The blood of chronic drunkards, reeking of alcohol and laced with impurities—"rawe & vnconcocted humors"—produced corrupted seed and therefore dysfunctional, handicapped, or alcoholic children. Robert Burton states emphatically that "Foolish, drunken, or hairbrain women most part bring forth children like unto themselves," and "if a drunken man get a child, it will never likely have a good brain ...."[46]

Although a tendency toward drunkenness usually was inherited from the parents, it could also be acquired from a hard-drinking nurse by way of breast milk, which was a derivative of feminine blood.

Burton thus mentions "Tiberius, who was a common drunkard, because his nurse was such a one."[47] King James I of England attributed his own habitually excessive drinking to his childhood ingestion of the milk of his nurse, Helen Littel, whose consumption of alcohol included at least one pint of wine and one gallon of ale—often more—per day.[48]

The hereditary effects of drinking had social and political implications as well. During the Renaissance, when class distinctions were based upon familial bloodlines,[49] alcoholic beverages had distinct class associations.[50] Good drink (i.e., wine) begot good blood, and good blood—hot and moist—produced robust health, sound judgment, courage, honesty, amiability, and other traits that were supposed to fit a human being out for entitlement to high social rank and responsibility. In theory, therefore, the complexions of the aristocracy—enriched by generations of blood kept hot by reason of access to the best wines, and carefully preserved and nurtured by other measures such as education, the strenuous cultivation and exercise of martial and civil virtues, and carefully arranged marriages—were sanguine, or nearly so, and these complexions were bestowed from one generation to the next by means of parental seed, which was derived from parental blood.

Baseborns, on the other hand, were supposed to be the inheritors of seed-blood that was cold and clammy. Themselves unable to afford wines, as were their parents, they had to make do with beers and ales[51]—beverages medically regarded as producers of thick, cold blood. Congenitally melancholic, baseborns were subject to behavior and states of mind arising from bodily coldness: at their worst, baseborns were cruel, treacherous, mean-spirited, dull, cowardly, and envious; at their best, they were agreeable but clownish, manipulative but naive, pretentious but ill-mannered. Forced to subsist upon miserable fare, especially in regard to drink, baseborns produced seed-blood that begot children like themselves: specimens of humanity inferior in practically every way to the offspring produced by the better-nourished aristocracy.

With society thus rigidly stratified according to the supposed quality of familial blood, there was little room for social mobility, regardless of the potential, such as it was, for members of the lower class to accumulate financial wealth. The assimilation of wealthy merchants into the aristocracy was a process that took generations to accomplish—and even then these "gentlemen of wealth" were generally looked upon as parvenus whose social status was unlegitimized by ancestral blood. It is upon this Renaissance consideration that Jack Cade, the rebellious peasant in Shakespeare's *Henry VI, Part II*, proclaims that when he becomes England's king he "will make it felony to drink small beer" and will "command that, of the city's cost, the pissing conduit run nothing but claret wine this first year of our reign" (IV.ii.72–73; IV.iv.2–5). Obviously Cade, who intends to level English society by elevating the peasantry to the peerage, plans to enact legislation to ensure that the blood of his new aristocracy is as sanguine as that of the old.

The most sanguine of all complexions (in theory, at least) was that possessed by the sovereign. Respect for the blood of the upper classes found its most poignant expression in descriptions of the blood royal, enriched by generations of sanguine ancestors and sustained by the best wines. A perfect balance of humors endowed the ideal monarch with a wide range of emotions and with degrees of mental and physical soundness that few human beings could hope to attain. "To enjoy a perfection in all the powers which governe man," writes John Huarte, "namely, the generative, nutritive, wrathfull, and reasonable is more necessary in a king, than in anie artist whatsoever."[52] The occultation of royal blood encompassed the passing on of the crown to lineal successors through the metaphysical body of the king—that is, the supernatural body as distinct from the purely physical body—which consisted largely in the privileges and rights accruing to it by virtue of ancestry and divine right. This ethereal body was passed on lineally to descendants through the power of the blood.[53] Hence, in claiming that "Blood hath ... the power to inherite Soveraigntie," Francis Markham says that

> *the Great and Puissant Estate of Borne-Kings, [is] a Raunge so sacred and beloued both of God and all Good men, that not any whatsoeuer can bee more accomplisht, or more excellent; and indeed infinite are the Reasons why these before all others should principally excell: for first of all, in themselues, and in their naturall, perfit, and predestinate creation, doubtlesse the most powerfull hand of the Almightie hath a more super-naturall, and diuine manner of working, then in the lower and more inferiour Bodies: so that without faile, all Excellencies, all perfections, and all Vertues both Diuine*

> *and Morall, finde in their Bosomes and Apprehensions, a fairer and goodlier Palace, earlier ready, richer furnished, and better both for their peace and safetie than can bee in any second Descent, or meaner Qualitie.*[54]

Such was the condition of blood in its highest, purest form.

Yet good blood was susceptible to degeneration. Members of the aristocracy who through habitual drunkenness corrupted their blood and impaired their faculties—and who, accordingly, begot children like themselves—jeopardized their own social standing and, indeed, the stability of the commonwealth itself. James Hart laments:

> *It was againe wont to be a proverbial speech, Drunke like a beggar: but now many of the Gentry (alas the more is the pitty!) doe assume this as a prerogative to grace their gentility. And what a pitty is it now, to see Gentlemen of faire estates, of antient houses, descended of noble parentage and pedigree, so farre to wrong themselves, as in Tavernes and Tap-houses to become a companion to any base varlet, swill-bowle, tosse-pot and pot companion?*[55]

A drunken aristocracy, as Hart implies, would mean a breakdown of the social order: earls, dukes, and even kings who corrupted their blood with habitually excessive drinking were transforming themselves, and any offspring they might produce, into creatures of inferior blood and ability. The matter thus ranges beyond what would appear, at least to twentieth-century minds, to be simply a matter of the aristocracy's besmirching its reputation through carousal with the lower order: it encompasses as well the idea of the literal transformation of the aristocrat into the baseborn through the corruption of good blood as the result of surfeit and other practices ranging from the unseemly to the vile. The impact of a degenerate aristocracy on affairs of state would of course be resoundingly negative. It was therefore necessary for statesmen and political writers at times to try to preserve respect for, and obedience to, a degenerate sovereign by invoking the notion of "the king's two bodies"—the belief that a king possesses both a physical body, subject to all the frailties and afflictions of the flesh, and a metaphysical body, vested with the royal authority bestowed by God. The concept of the king's two bodies made it possible to distinguish the man from the monarch: physical and moral shortcomings could be attributed to the physical body, leaving the metaphysical body unblemished. The concept was repeatedly applied to King James I as a means of preserving at least minimal respect for this otherwise poor specimen of royal blood whose drunkenness, homosexuality, and general inattention to duty would else have undermined the authority of the crown.[56]

The association—indeed, the equation— of drink with blood was sufficiently pervasive during the Renaissance to find its way into the popular literature of the time, where it flourished alongside other scientific ideas—especially the association of blood with personality, behavior, and class-worthiness—as the stuff of images and allusions, nuancings and motivation. Shakespeare's plays, for instance, are rich in the use of all these ideas. Falstaff knows that sack rids the brain of vapors, sharpens the intellect, warms the blood, and rouses the vital spirits, whereas thin drink cools the blood (*Henry IV, Part II* IV.iii.96–201). Wine heats the liver, as Gratiano tells Antonio (*The Merchant of Venice* I.i.81), and a hot, red liver is the source of courage (*Richard III* IV.iv.465; cf. *King Lear* IV.ii.50) as well as sexual appetite and carnal vitality (*Love's Labors Lost* IV.iii.74–75; *The Tempest* IV.i.56). A liver deprived of veinous blood (*Twelfth Night* III.iii.65–66), or one that is no larger than a pigeon's (*Hamlet* II.ii.605), renders its possessor a coward (*The Merchant of Venice* III.ii.86; *Henry V* III.ii.33–35), whereas a liver heated with wine suffuses the body with spirits that inspire courage and martial vigor (*Antony and Cleopatra* I.ii.23).

Wine's heat is efficacious in the treatment of chilly agues, as Stephano drunkenly notes while urging more wine on Caliban (*The Tempest* II.ii.78–79, 97–98). It prolongs life by replenishing the supply of blood, as Benedick observes (*Much Ado about Nothing* I.i. 252–56), so the dying Antony calls for it as a means of keeping himself alive long enough to finish what he has to say (*Antony and Cleopatra* IV.xv.42). The toast Simonides drinks to Pericles is interpreted by Thaisa as Simonides' way of wishing more blood, and so a longer life, into Pericles' body (*Pericles* II.iii.77).

The effect of "hot" drinks on choleric personalities underlies Iago's urging a second cup of wine on Cassio, a choleric who has "very poor and unhappy brains for drinking" (*Othello* II.iii.34–35). Iago knows that more wine will add more heat to Cassio's naturally choleric blood, rendering him susceptible to a fit of anger that Iago can manipulate to his own ends (II.iii.50–53). A man could even become "red-hot with drinking," as Ariel facetiously observes apropos of Caliban and Trinculo

(*The Tempest* IV.i.171). The excessive drinking of wine could so overheat the blood as to impair or destroy the rational faculties, leaving the drinker in a condition that is monstrous (*Antony and Cleopatra* II.vii.104–6) or beastly (*Othello* II.iii.290–94) or, at best, foolish (*Hamlet* I.iv.19–22). Drunkenness incites hypocrisy (*Timon of Athens* I.ii.140–44), as well as swaggering and swearing (*Othello* II.iii.281). Lechery "it provokes and unprovokes" (*Macbeth* I.iii.32)—that is, it heats the blood and so triggers the inclination, but it takes away the physical capacity to perform.

Yet "good wine is a good familiar creature, if it be well used" (*Othello* II.iii.313–14), and in Shakespeare's plays, as in Renaissance society, its benefits as a purveyor of amiability, courage, strength, and long life—all properties arising from bodily heat—make it the preferred beverage of the aristocracy. Nourished by wines, the blood of Shakespearean gentlemen is characteristically hot, and its heat emerges in a tendency toward choler. "That which in mean men we entitle patience," the Duchess of Gloucester tells Gaunt, "Is pale cold cowardice in noble breasts" (*Richard II* I.ii.33–34). The "gentleman of blood" (*Two Gentlemen of Verona* III.i.21) in Shakespeare's plays is expected to engage "sprightfully and bold" in trials by combat and to face all occasions of mortal danger "lustily ... and cheerly drawing breath" (*Richard II* I.iii.3, 66). He is supposed to be war-like (*Henry IV, Part I* V.iv.18; *King John* II.i.350–51). He disdains "calm, dishonorable, vile submission" when provoked by disparagement of himself or his family (*Romeo and Juliet* III.i.76) because the "high blood chafed" by insult or injury (*Troilus and Cressida* Prologue line 2) can often be assuaged only by bathing in the heart-blood of the offender (*Richard II* I.i.172; *Henry IV, Part I* III.ii.135–36). Yet the gentleman of blood confronts his adversaries with modest rather than bombastic or rancorous challenges (*King Lear* V.iii.129–41; *Henry IV, Part I* V.ii.52–65). He is chivalrous to those whom he has defeated in battle (*Troilus and Cressida* IV.v.185–91; V.iii.40–42). But he is not unduly sanguine. Even a king whose blood is "too cold and temperate" can find himself on the shaky throne of a kingdom ravaged by insurrection (*Henry IV, Part I* I.iii.1–2).

The choler of the ideal Shakespearean gentleman is nevertheless tempered by the rich moisture that wine imparts to humoral blood. Such a character is not, as for instance Hotspur is, so choleric as to be "harebrained" (*Henry IV, Part I* V.ii.19). His sanguinity tempers his choler and thereby endows him with patience in the face of adversity (*Coriolanus* IV.i.7–9) as well as firm resolution in his moral and ethical choices despite strong opposition or adverse circumstances (*Henry IV, Part I* IV.i.38).

Though predominantly hot, the blood of even the best Shakespearean gentles has its share of the cold humors and is therefore not exempt from carrying such "cold" passions as sorrow and grief (e.g., *Henry IV, Part II* V.ii.49–53; *Pericles* IV.i.24). The presence in good blood of all four of the humors endows the noble soul with a high capacity for experiencing the gamut of human feeling. A gentleman is therefore subject to compassion as well as wrath, to choler as well as sorrow and fear (*Henry IV, Part II* IV.iii.31–41, V.ii.49–53; *Julius Caesar* V.v.73–75; *Antony and Cleopatra* V.ii.82–89). Martial courage coexists alongside the "milk of human kindness" in noble blood (*Macbeth* I.v.18). Thus Romeo's blood, rendered "effeminate" by Juliet's beauty, turns instantly choleric when he catches sight of Tybalt (*Romeo and Juliet* III.i.118ff.).

In fact, Shakespeare's gentle characters may experience fear or grief so deeply that their hearts break, freeze, or explode, as happens to Gloucester and the King in *King Lear* as well as to such other Shakespearean gentles as Brabantio, Enobarbus, Mamillius, and Sicilius Leonatus. Such deaths are reserved only for Shakespeare's gently born characters: no baseborn in the plays ever dies of a broken heart.[57] The inability to suffer a broken heart is something the baseborn Parolles regrets: "If my heart were great," he sighs, "'Twould break at this" (*All's Well That Ends Well* IV.iii.366–67). Parolles, by reason of his sluggish blood, cannot experience passion to so considerable a degree. Gentles also experience fear (e.g., *Hamlet* I.v.16), though presumably this would not be the "vassal fear" with which Henry IV believes his apparently degenerate son to have been somehow endowed (*Henry IV, Part I* III.ii.124) and which is associated with the cowardice endemic to Shakespearean baseborns.

Understandably, Shakespearean gentles usually strive to attain a rational control over their passions, as for instance Prospero does in *The Tempest* V.i.26–28, because to act rashly or to behave in some otherwise indecorous manner is ignoble (*King John* II.i.48–49). Honor cannot be divorced from wisdom and rational behavior (*Henry IV, Part II* I.i.162).

Because their blood is rich and hot, many of the gentle characters in Shakespeare's plays display an uncanny facility for learning. Their minds seem naturally to "draw in / More than the common blocks" (*The Winter's Tale* I.ii.224–25). Much of what they know and many of the talents they possess appear not to have been matters of acquisition or development but to have been things born into them—evidence of high quality seed-blood, nourished by wines, that has been inherited from parents and forebears. Camillo says of Perdita, a gentlewoman by birth who has been reared by shepherds, "I cannot say 'tis pity / She lacks instructions, for she seems a mistress / To most that teach," and Florizel says that "She is as forward of her breeding as / She is i' the rear o' her [apparent] birth" (*The Winter's Tale* IV.iv.591–93; IV.iv.590–91). The two sons of Cymbeline, though reared in the wilderness and therefore bereft of formal schooling, possess a keenness of intellect and a degree of innate knowledge incompatible with their base circumstances (*Cymbeline* IV.ii.176ff.). Orlando, though denied the education to which his birth entitles him, nonetheless is "never schooled, and yet learned" (*As You Like It* I.i.172–73).

Shakespearean gentles who have the benefit of formal instruction are usually fast, effortless learners. Posthumus, for instance, was exposed to "all the learnings that his time / Could make him the receiver of, which he took / As we do air, fast as 'twas administered" (*Cymbeline* I.i.43–45). Prince Hal, though surrounded by "companies unlettered, rude, and shallow," suddenly becomes, to the amazement of Canterbury, a prodigious scholar. "Never was such a sudden scholar made," Canterbury says (*Henry V* I.i.32). Canterbury's remark that "Consideration like an angel came / And whipped the offending Adam out of [Hal], / Leaving his body as a paradise / To envelop and contain celestial spirits" (I.i.28–31) may be an allusion to animal spirits, the "celestial" substances derived from blood that were responsible for intellection and moral judgment and that were also regarded as the locus of the soul (cf. *Love's Labors Lost* IV.iii.306). At any rate, good blood in Shakespeare's plays generally means formidable intellectual powers which, when nurtured through instruction in "arts and arms" (*Pericles* II.iii.82), result in the "noble mind" of the complete gentleman, who is at once a courtier, a soldier, and a scholar (*Hamlet* III.i.58–59).

Inferentially, most of Shakespeare's gentles are schooled in music and letters (*Pericles* II.v.30), mathematics, poetry, and languages (*The Taming of the Shrew* I.i. and I.ii. passim), and the graces of courtly behavior (*Pericles* IV Chorus lines 9–10). This curriculum would presumably be beyond the mastery of baseborns, whose diminished spirits (*Antony and Cleopatra* V.ii.157) make learning all but impossible (*Henry V* I.i.28–31). Baseborns are like Caliban, who, despite Prospero's attempts to elevate him intellectually and morally through instruction, remains a creature "on whose nature / Nurture can never stick" (*The Tempest* I.ii.351–52). They are like the degenerate knight Aguecheek, whose poverty of blood (*Twelfth Night* III.ii.66) betokens not only deficiency in courage but also a poor capacity for learning, as well as a dearth of natural talent (I.iii.28ff., 98–99).

Properly nurtured by attention to diet and by the exercise of moral and civil virtues, gentle blood is supposed to make its possessors exceptionally honest, courageous, and amiable (*Henry IV, Part I* I.ii.157–59) as well as almost preternaturally keen and discerning (*The Winter's Tale* I.ii.222–28). But gentle blood in Shakespeare's plays is susceptible to degeneration or depletion. It can be "bred out" (*Henry V* III.v.29), or it can be corrupted by the cultivation of vile thoughts and morally base desires (*Twelfth Night* III.iv.390). Even the blood royal can be "disfigured clean" by aberrant sexual practices (*Richard II* III.i.9–14). Its pristine condition can be mitigated by repeated behavior that is morally contrary to the blood's natural inclination (*Coriolanus* III.ii.120–30; *Hamlet* III.iv.168). It can be vitiated by food and drink of poor quality (*Henry IV, Part II* II.ii.11–13).

Even the best blood was liable to the degenerative effects of a poor or improper diet because the consumption of inferior food and drink could engender veinous blood heavily tainted with melancholy. For instance, following his exile from Athens, the noble Timon appears to subsist exclusively on roots and cold water (*Timon of Athens* IV.iii.s.d.; V.i.77), a diet that no doubt contributes much to the melancholic frenzy that consumes him after his banishment, for as Gloucester observes in I.i.139–40 of *Richard III*, a poor diet breeds melancholic blood that overthrows the finer powers of mind and body. The corruption as well as the unwholesome chilling of good blood with bad food and drink is the point behind Cloten's denunciation of Posthumus as "One … fostered with cold dishes, / With scraps o' the Court ..." (*Cymbeline* II.iii.19–20). The deliberate corruption of blood by means of a poor diet figures in the punishment Prospero threatens to impose on Ferdinand: the latter is to be imprisoned, says Prospero, and fed with "The fresh-brook mussels, withered roots, and husks / Wherein the acorn cradled" (*The Tempest* I.ii.463–64). Such a diet would prove not only distasteful but also detrimental to the quality of Ferdinand's aristocratic blood, leaving him both physically weaker and subject to the psychological afflictions of severe melancholy.

If Shakespeare's plays often testify to the high quality of blood enriched by wines, they also reflect something of the poverty of blood produced by the drinking of beers and ales. In *Henry VI, Part II,* for example, Jack Cade promises to make the consumption of small beer a felony (IV.ii.73), presumably because small beer—a producer of cold blood—would preserve the melancholic complexions of the baseborns whom he intends to elevate to the peerage. In *The Taming of the Shrew,* the rustic tinker Christopher Sly rejects a cup of sack in favor of "a pot of small ale" (Induction ii.1ff.)—another beverage thought to produce cold, sluggish blood. Shakespeare thus links beers and ales with brown bread and garlic (*Measure for Measure* III.ii.193–94), toasted cheese (*Henry VI, Part II* IV.vii.13), porridge (*Troilus and Cressida* I.ii.263), onions (*A Midsummer Night's Dream* IV.ii.42–43), and other fare as staples of the lower-class diet. From such blood as these staples were thought to produce come the repulsive, melancholic traits of so many Shakespearean baseborns: cowardice (*Henry VI, Part II* III.i.33–36) and "vassal fear" (*Henry IV, Part I* III.ii.124), hard-heartedness, fickleness, stupidity, and cruelty (*Julius Caesar* I.i.41–56). As the inheritors of poor blood who sustain themselves on miserable food and drink, baseborns are "The mutable, rank-scented many" capable of displaying courage only when they are engaged in mutiny or revolt (*Coriolanus* III.i.66, 125–27). They are "idle creatures," "blocks," "stones," "worse than senseless things" who possess dry, chapped skin; who sweat profusely; and who befoul the air with a remarkable halitosis (*Julius Caesar* I.i.1; I.ii.245–47). They are untrustworthy, envious, spiteful, and gossipy (*Coriolanus* I.i.168–207). They are dissolute rebels motivated by avarice and fear, the passions that arise naturally from cold, melancholic blood (*King John* IV.iii.141–52).

Impoverished by beers, ales, and other poor fare, base blood leaves the body cold, and a cold body means sterility (*Henry VI, Part II* II.iv.3) and short life (*Hamlet* I.v.16). Base blood can be ancestral only in the narrowest sense, inasmuch as baseborns must necessarily have biological antecedents (*Coriolanus* II.i.103–4). Yet so poor is the quality of the blood inherited from parents and forebears that Shakespearean baseborns are, in effect, bastards.[58] As Caius says of the Roman peasantry, baseborns are "fragments" (*Coriolanus* I.i.226)—incomplete human beings who possess so little humoral blood that their organs cannot produce in quantity the animal spirits by means of which the moral and intellectual faculties are empowered. Accordingly, baseborns may be said to lack human souls (*Antony and Cleopatra* V.ii.157). Lacking spirits, baseborns operate—like beasts—at the emotional and sensory rather than the moral and intellectual levels (*Hamlet* IV.iii.5). Baseborns therefore have the souls of geese and camels (*Coriolanus* I.iv.34; II.i.266–67).

For all his harshness in rendering the poor characteristics of baseborns, Shakespeare nevertheless presents us with some such characters who are redeemed from the charges he generally levels at their caste. In *King Lear*, for instance, the baseborn servant of Cornwall comes to the defense of the helpless Gloucester, an act of uncharacteristic heroism for which he pays with his life. Pisanio in *Cymbeline*, Adam in *As You Like It*, and Flavius in *Timon of Athens* are among the few other baseborn servants whose honesty, steadfastness, devotion, and generosity are atypical of a class supposed to produce only creatures like the fawning, cowardly Oswald in *King Lear*. Even the mechanicals in *A Midsummer Night's Dream*, though obviously "common blocks" whose obtuseness is endemic to their baseborn nature, are redeemed by a certain charm that, to some degree, separates them from their mean-spirited counterparts in *Julius Caesar*, *Coriolanus*, *Antony and Cleopatra*, and other plays. These characters lack the vileness of soul that Shakespeare imputes to other baseborns.

No doubt these differences are attributable in part to the differences in the nature and the purposes of the plays themselves. Nevertheless, the misperceptions, the manglings of language, and the other traits that are the hallmarks of Bottom and his companions serve as continual reminders of their station and their blood. They are as far from being gentles as lead is from being gold. In matters of class consciousness, it is as the King of France says in *All's Well That Ends Well*: "Strange is it that our bloods, / Of color, weight, and heat, poured all together, / Would quite confound distinction, yet stand off / In differences so mighty" (II.iii.125–28).

These differences among Shakespeare's characters—differences in longevity, vigor, personality, intelligence, and social station—are the particular manifestations of blood that has been impoverished or enriched, usually over generations, by drink. The association of drink with blood, like the identification of blood with heredity and with the behavioral and moral impulses in human nature, is something that turns up frequently not only in the works of Shakespeare but also in the poems and plays of virtually all Shakespeare's contemporaries or near-contemporaries: Jonson, Donne, Herbert, Chapman, Ford, and others routinely use it in matters of imagery and characterization. Although its imagistic significance, its bearing on characterization, and its importance in regard to episodic or thematic concerns naturally may vary from author to author and from work to work by reason of authorial bias and purpose, the association—indeed, the equation—of drink with blood pervades the poems and plays of the time, and to understand that equation is to possess a lamp which brightens much in Renaissance literature that is otherwise obscure to modern minds.

Notes

1 Readers of this essay who wish to acquaint themselves more thoroughly with the general principles of Renaissance medical science will find a plethora of works on the subject. A few such works from which this essay silently draws are Ruth L. Anderson's *Elizabethan Psychology and Shakespeare's Plays* (New York: Russell & Russell, 1966), John W. Draper's *The Humors and Shakespeare's Characters* (Durham: Duke University Press, 1945), Lawrence Babb's *The Elizabethan Malady* (East Lansing: Michigan State College Press, 1951), and John Erskine Hankins' *Backgrounds of Shakespeare's Thought* (Hamden: Anchor Books, 1978).

2 Timothy Bright, *A Treatise of Melancholie* (1586; rpt. New York: Facsimile Text Society, 1940) 4–5.

3 Robert Burton, *The Anatomy of Melancholy*, ed. Holbrook Jackson (New York: Vintage Books, 1977) I: 147–48.

4 Thomas Walkington, *The Optick Glasse of Hvmors* (London, 1607) 65r.

5 Sir Thomas Elyot, *The Castell of Helth* (London, 1539) 29.

[6] Bright 28.

[7] James Hart, KLINIKE, *or the Diet of the Diseased* (London, 1633) 126.

[8] Burton I: 223.

[9] Hart 126.

[10] Hart 127.

[11] Hart 127

[12] Burton I: 223.

[13] Hart 128.

[14] Burton I: 153.

[15] Andre du Laurens, *A Discovrse of the Preseruation of the Sight,* trans. R. Surphlet (London, 1599) 170.

[16] Walkington 25v.

[17] Nicholas Gyer, *The English Phlebotomy* (London, 1592) 44.

[18] Walkington 25v.

[19] William Turner, *A New Boke of the Natures and Properties of All Wines* (London, 1588) B3v.

[20] Hart 120, 121.

[21] Thomas Thompson, *A Diet for a Drvnkard* (London, 1612) 33.

[22] Hart 127.

[23] Burton I: 223.

[24] Hart 123.

[25] Hart 112, 113.

[26] Hart 113.

[27] Roger Bacon, *This Boke Doth Treate All of the Beste Waters Artyfycialles* (np, ca. 1550) A2r–v.

[28] The continuing popularity of Anthonie's *aurum potabile* is verified by the celebrated Restoration chemist Robert Boyle in *Of the Reconcileableness of Specifick Medicines to the Corpuscular Philosophy* (London, 1685) 182–85.

[29] Francis Anthonie, *The Apologie, or Defense of a Verity Heretofore Pvblished Concerning a Medicine Called* Avrvm Potabile (London, 1616) 113.

[30] Anthonie 114.

[31] Anthonie 11–12.

[32] John Woodall, *The Cvre of the Plague by an Antidote Called* Avrvm Vitae (London, 1640) A4v.

[33] Woodall A4r.

34 Woodall B1v, B2r.

35 Woodall B2r.

36 Burton I: 154.

37 Gyer 67.

38 J. Oswen, *An Inuectyue agaynst Dronkennes* (Ippiswiche, 1548) A8r.

39 Hart 131–32.

40 Thomas Wright, *The Passions of the Minde in Generall* (London, 1621) 8.

41 Wright 8

42 Oswen A7r.

43 Hart 129–31.

44 Thompson 37.

45 Thomas Cogan, *The Haven of Health* (London, 1589) 240.

46 Burton I: 213.

47 Burton I: 331.

48 Henry N. Paul, *The Royal Play of Macbeth* (New York: Octagon, 1971) 389.

49 See, for instance, David Castronovo's *The English Gentleman: Images and Ideals in Literature and Society* (New York: Ungar, 1987), especially pages 5–30, and W. Lee Ustick's "Changing Ideals of Aristocratic Character and Conduct in Seventeenth-Century England," *MP* (1932): 147–66.

50 Gallobelgicus in *Wine, Beere, and Ale, together by the Eares,* trans. Mercurius Brittannicus (London, 1629), an allegorical treatise on the class associations of various beverages, presents Wine as "a Gentleman," Beer as "a Citizen," and Ale as "a Countrey-man." These characters, involved in a dispute as to which of them is the most sovereign replenisher of blood, are joined by Water, "a Parson," who resolves the dispute by showing that Wine is best for "Galants, Gentlemen, and Poeticall Wits"—those who are "of a refined mould" and who require "a more nimble and actiue watering, to make their braines fruitfull," whereas Beer and Ale are more suitable for citizens and rustics, respectively.

51 William Younger, *Gods, Men, and Wine* (Cleveland: World Publishing Co., 1966) 292.

52 John Huarte, Examen des Ingenios, *the Examination of Mens Wits,* trans. R. Carew (London, 1616) 252.

53 The concept of the king's two bodies, discussed later in this essay, is fully explored by Ernst H. Kantorowicz in *The King's Two Bodies: A Study in Medieval Political Theology* (Princeton: Princeton University Press, 1957).

[54] Francis Markham, *The Boke of Honovr, or Five Decads of Epistles of Honovr* (London, 1625) 181.

[55] Hart 135.

[56] Jonathan Goldberg, *James I and the Politics of Literature* (Baltimore: Johns Hopkins University Press, 1983) 40–54, 140–50. See also Kantorowicz, above.

[57] David Shelley Berkeley, *Blood Will Tell in Shakespeare's Plays* (Lubbock: Texas Tech Press, 1984) 24. In connection with this, I should like to point out that Marvin Spevack in "The Art of Dying in Shakespeare," *Shakespeare-Jahrbuch* (1989): 169–73, has shown that the nature of death in the plays varies according to the gender as well as the social class of the individual characters.

[58] John W. Draper in "Bastardy in Shakespeare's Plays," *Shakespeare-Jahrbuch* 74 (1939): 123–36, has demonstrated that Shakespeare's lower-class characters share many of the traits of his bastards.

The Fool as Physician in Shakespeare's Plays

by Kenneth J. Tiller

> *A merry heart doeth good*
> *like a medicine: but a broken*
> *Spirit drieth the bones* (Proverbs 17. 22)

The Renaissance fool, in both society and literature, has been a traditionally curious subject. Intrigued by the relationship of the often incredibly impudent professional jester and the increasingly autocratic rulers of that period, scholars, most notably Enid Welsford and William Empson, have discussed at length the various facets of this character's role in Renaissance society and in Shakespearean drama.[1] Without slighting the contributions of these important commentators, I will emphasize an aspect of the fool's duty that seems to have been neglected: the therapeutic virtue of making his or her liege laugh. Although this aspect of the fool's function may seem too obvious to warrant commentary, reexamination of Renaissance medical practices, with their close interrelationship between psychological and physiological health, suggests a previously overlooked medicinal function of foolery. This function, evident in Renaissance tracts on both medicine and foolery, is reflected to a significant degree in the plays of Shakespeare.

Although previous scholarship has noted the fool as a restorative agent, critics tend to base their work on modern psychological discourse and therefore to overlook the physiological aspects of foolery.[2] For example, Brian A. Spillane, seeing Lear's fool as a representative of balance and unity and hence of a "healthy" political and physical condition, lays important groundwork for a study of the medicinal value of the fool: "We find that the fool himself is a figure of head and heart, acting out of reason and emotion in a balanced way that would mitigate the unnatural divisions of the play."[3] However, grounding his research in philosophy and modern psychotherapy, Spillane does not address the physiological-psychological forms of therapy actually practiced in Shakespeare's period. James Black, arguing that the fool "must speak for life in the face of morbidity,"[4] perhaps comes closer to appreciating this character's health-restorative virtues. Neither Spillane nor Black, however, considers Shakespeare's fools in the context of Renaissance medicinal writings on sorrow, melancholy, mirth and foolery.

From the perspective of Renaissance medicine, with its emphasis on the physiological effects of passions and humors, a fool, being more than an amusing entertainer, could play an indispensable role in the maintenance of his or her liege's health. In view of the belief that the mirth provided by fools was thought to have a direct restorative effect on the health of those being entertained, I propose that Shakespeare's presentations of the fool, employing the medicinal virtues of mirth, display the dominant medical views concerning the passions and humors. Beginning with an examination of Renaissance medical views of melancholy and sorrow and their physiological consequences, I will first show that Renaissance treatises suggest mirth and foolery as a remedy. Having established a context for an examination of foolery as therapy, my next section will point to Shakespeare's engagement with medical practices of his period concerning sorrow, melancholy, and mirth and foolery. Maintaining that the fool's medicinal role is one that Shakespeare exploits fully in his deployment of images of health and medicine, I will then provide necessarily brief analyses of five plays—*The Comedy of Errors*, *All's Well That Ends Well*, *As You Like It*, *Twelfth Night*, and *King Lear*—in which the fool figures prominently. My study will conclude by arguing that the Shakespearean fool is, in a sense, a "physician" skilled in relieving both individuals and communities of the degenerative effects of sorrow and melancholy.

Elizabethan Medical Views of Sorrow, Melancholy, Mirth and Foolery

"The Elizabethans," Hardin Craig has observed, "were by and large more attentive to the physiological effects of passion than we are."[5] In the Renaissance, passions were considered capable of causing the heart either to expand (as in the case of joy) or to contract (during sorrow or mourning). They directly altered the flow of the four bodily humors: blood (warm and moist), choler (warm and dry), phlegm (cold and moist) and melancholy (cold and dry). These four disparate substances present in the blood likewise referred both to physical substances and to psychological conditions.[6] Although they eventually took on strictly psychological meanings, at least one of these humors—melancholy—retained its physiological implications well into the eighteenth century.[7] Therefore, an Elizabethan physician, practicing holistic medicine and accustomed to thinking of changes in mood as being associated with imbalances of bodily humors, would have no need for the term "psychosomatic illness." *All* diseases, related to various passions and humors, would be, in a sense, psychosomatic.

A survey of Elizabethan medical views reveals a great many physiological maladies attributed to sorrow and melancholy. Sorrow, originating in the heart, was capable of contracting the heart, stifling the flow of blood, and depriving the body of much-needed natural heat. William Vaughan, a major medical writer of the period, writes that "sorrow stifleth up the purere faculties of the soule, causeth a man to fall into a consumption, and to be weary of the world, yea, and of himself."[8] Sir Thomas Elyot likewise cites sorrow as a physically as well as psychologically detrimental condition; specifically, it eroded the vital heat and moisture considered necessary to life and health: "There is nothynge more ennemye to lyfe, than sorowe, callyd also hevynes, for it exhausteth bothe naturall heate and moysture of the bodye" (63–4). Shakespeare, using physiologically charged terms such as "passions," "care," and "melancholy" to present the debilitating effects of sorrow in his plays, reveals his cognizance of medical wisdom. In *Twelfth Night*, Sir Toby condemns his niece Olivia's seven-year period of mourning for her brother: "I am sure care's an enemy to life" (I.iii.2–3), a sentiment Lafew echoes in *All's Well That Ends Well*, advising Helena that "moderate lamentation is the right of the dead, excessive grief the enemy of the living" (I.i.64–65). Further, the passion was particularly dangerous because of its ability to corrupt and diminish the blood supply (and Professor Berkeley has aptly demonstrated the importance Shakespeare placed on maintaining the integrity of the blood), an effect of which might be the degeneration of an otherwise young, healthy individual into a state resembling advanced age. Cymbeline, for instance, threatens that his daughter may "languish / A drop of blood a day, and being aged / Die of this folly!" (*Cymbeline* I.i.154–56).

The physiological danger of sorrow was further exacerbated by its relationship to melancholy, the thick, dark, semi-excremental substance that became a near obsession of both Elizabethan physicians and poets. Renaissance physicians recognized two or three sorts of melancholy: "natural" melancholy, an overabundance of the humor brought about by a variety of physical or psychological causes, was a dark, heavy substance capable of corrupting the blood and sending noxious fumes into the brain. Although this humor was not always destructive—"Pseudo-Aristotelian" or "Witty" melancholy, a moderate amount of the humor mixed with otherwise high quality blood, was considered desirable and was often affected by courtiers—melancholy was usually treated as a malady. As Timothy Bright says, the humor consists of

> *the grosser part of the blood ordained for nourishment, which either by abundance or immoderate hotenesse, passing measure, surchargeth the bodie, and yeldeth up to the braine certaine vapors, whereby the understanding is obscured; or else is an excrement ordained to be avoyded out of the bodie ... having not a drop of nourishing iuyce remaining, whereby the bodie, either in power or substance may be relieved.* (1–2)

The results of excessive melancholy were sluggishness, cowardliness, and lethargy. A second, more dangerous form, unnatural or adust melancholy, was formed by burned humors (either blood, melancholy, or choler), "whollie chaunged into another nature by an unkindly heate, which turneth these humours, which before were raunged under natures government, and kept in order, into a qualitie whollie repugnant, whose substance and vapor giveth such annoyance to all the partes, that as it passeth or is seated maketh strange alterations in our actions." Whereas the main symptoms of the first sort of melancholy were lethargy and cowardice, this form of melancholia engendered insanity and often death.

Manifestations of the melancholy humor provided a popular subject for dramatists as well as for physiologists. Ben Jonson, for instance, mocks the gentry's affectations of witty, or pseudo-Aristotelian, melancholy in his *Every Man in His Humour.* Shakespeare, less satiric in his treatment of the subject, seems more attuned to the serious physiological consequences of melancholy. In *The Comedy of*

Errors, the abbess warns Adriana that her jealous clamoring will lead to melancholy and to the consequent ruin of her husband's health:

> *Sweet recreation barr'd, what cause doth ensue*
> *But moody and dull melancholy,*
> *Kinsman to grim and comfortless despair?*
> *And at her heels a huge infectious troop*
> *Of pale distemperatures and foes to life?* (V.i.78–82)

Here, the abbess recites the inescapable association between sorrowful emotions, the engendering of the melancholy humor, and detrimental effects on the health. Adriana's earlier observation that her husband was "heavy" and afflicted with "passions" indicates clinical conditions, in this case to be treated with "wholesome syrups, drugs, and holy prayers" (V.i.104). Similarly, Timon's madness, especially his solitariness, is attributed to an "infection" brought about by melancholia: "This is in thee a nature but infected, / A poor unmanly melancholy sprung / From change of future" (*Timon of Athens* IV.iii.202–4). In addition to madness, melancholy could lead to death. Timon probably dies of melancholy, and the "green and yellow" illness afflicting Viola's (fictional) sister in *Twelfth Night* seems at least potentially deadly. Therefore, although occasionally satirizing the gentry's melancholic pretensions—as he does with Jaques's cynical pretensions (*As You Like It*) and Romeo's "love melancholy" (*Romeo and Juliet*)—Shakespeare generally seems more concerned than does Jonson with the physiologically destructive aspects of the condition.

If sorrow and melancholy caused physical deterioration by cooling, thickening, and reducing the blood supply, then, according to the Galenic principle of curing by opposition, mirth could effectively restore the blood's desired warmth and thinness by enlarging the heart and sending forth warmer blood in greater supply. Indeed, laughter and jovial company almost always figure prominently among the regimen physicians prescribed against melancholy. John Makliure, in *The Buckler of Bodilie Healthe*, asserts the heat-restorative powers of mirth: "it [mirth] sendeth forth an abundance of the naturall heate with blood" (60). Edward Reynolds adds that mirth acts "to dispell all adverse humours [especially, no doubt, melancholy] that would hinder the ingresse of [spirit]" (218). Burton devotes an entire chapter of his famous *Anatomy of Melancholy* to the therapeutic value of mirth: "Mirth ... purgeth the blood, confirms health, causeth a fresh, pleasing, and fine colour, prorogues life, whets the wit, makes the body young, lively, and fit for any manner of employment" (481). Significantly, Burton explicitly cites fools and jesters as agents against melancholy: "Princes use jesters, players, and have those masters of revels in their Courts" (481).

In light of the virtues attributed to mirth, fools could have immense therapeutic value. Indeed, contemporary writers on foolery echo Burton in referring to the fool's ability to cure melancholy. Jestbooks advertised themselves as remedies for melancholy, a practice that continued into the eighteenth century.[9] Robert Armin, Shakespeare's famous stage fool (he created the roles of Touchstone, Feste, and Lear's fool) refers to the fool as a "please-plaster for melancholy."[10] In his jestbook *A Nest of Ninnies*, he relates how Henry VIII was once cured of melancholy by Will Somers, his court fool: "The King upon a time being extreame melancholy and full of passion, all that Will would doe will not make him merry. 'Ah,' says hee, 'this [cloud] must have a good showre to cleanse it,' and with that goes behind the Arras. 'Harry,' says he, 'I'll go behind the arras and study three questions and come again. See therefore you lay aside this melancholy muse and study to answer me." [11] Although Will's riddles seem to the modern reader more scatological than amusing, Armin's anecdote, depicting a fool tending to the physiological-psychological well-being of his liege, confirms the medicinal value of foolery. Further, Renaissance nobles themselves seemed to have recognized therapeutic practices as part of their kept fools' duties. Welsford mentions more than one instance of a professional fool curing or relieving illness: "In 1498, Alfonso d'Este fell seriously ill, and Isabella sent her favorite fool Matello to cheer him. His success was amazing. Alfonso writes of Matello in most enthusiastic terms, saying that he made him forget the severity of his illness."[12] Although Welsford notes with some surprise that the fool "seems oddly enough to have been particularly welcome in the sick room,"[13] tending the sick would, to Alfonso and his contemporaries, as well as to Shakespeare, be seen as part of a professional fool's duties.

Shakespeare's Fools as Physicians

While Shakespearean scholarship has acknowledged the healthful benefits of music, sport, and rest, it has tended to overlook the equally valuable medicinal benefits of mirth, which, as evidence indicates, was recognized throughout the Renaissance and beyond as a cure for melancholy and for sorrow-associated illnesses.[14] I do not, of course, suggest this function to be the only duty of kept fools; I do, however, believe that an understanding of the fool's medicinal value may help clarify the often complex relationship of this character to his liege, and, further, that it reveals an added dimension to Shakespeare's use of medical knowledge and the language of medicine in the plays, especially regarding the association of the fool and the melancholic. Examination of Shakespeare's use of medical terms relating to melancholy and sorrow, and the well-established value of mirth as a health preservative, suggests that Shakespeare's fools are, in a sense, "physicians," specializing in the purging of melancholy and the relief of sorrow.

In *The Comedy of Errors*—Shakespeare's first comedy—Antipholus of Syracuse, in a short explanation echoing Renaissance medicine, lists medicinal jesting as one of his servant Dromio's main functions:

A trusty villain, sir, that very oft,
When I am dull with care and melancholy,
Lightens my humor with his merry jests. (I.ii.19–21)[15]

Remembering that terms such as "dull," "care," and "melancholy" referred to physiological as well as to psychological conditions, this passage credits Dromio with much more than occasionally "cheering up" Antipholus. Rather, the occasional melancholy and attendant "dullness" of which Antipholus complains would be interpreted as a literal imbalance of humors, an excess of black bile that, if unchecked, could cripple Antipholus's health and sanity. Likewise, the power of Dromio's jesting to "lighten his humor" would refer to the very literal thinning and lightening —in terms of both weight and color—of Antipholus's blood and the purgation of excess melancholy. Significantly, Antipholus indicates that the relieving of melancholy by means of jesting is an officially recognized part of Dromio's duties; his servant and sometimes fool, therefore, constitutes part of his regimen of personal health. Shakespeare thus exhibits, relatively early in his career, familiarity with his period's views of melancholy and the healthful benefits of mirth.

Dromio's presence itself enhances the medicinal theme of this early farce, complementing the "food, sport and life preserving rest" (V.i.83) recommended for the maintenance of bodily health.[16] Adding an element of false madness to the Plautine farce, Shakespeare could well enhance comic irony of the situation by having the Antipholus brothers employ professional fools, an acknowledged preventative against melancholy; in this way their actual sanity would be underscored. The relationship between Antipholus and his fool-servant, furthermore, foreshadows the more sophisticated fool-melancholia associations evident in the later plays.

In Shakespeare's later comedies, fools figure most prominently in situations where melancholy and melancholic humors threaten the comic festiveness. Lavatch, the Countess of Rossillion's fool in *All's Well*, appears in a realm characterized by sorrow and nostalgia (passions capable of engendering melancholy): Helena mourns immoderately for her father, and the Countess mourns for her son Bertram's departure for the Italian war.[17] Their sorrowing is linked, by the theme of disease, to the French king's actual fistula and to the language of disease evident in his courtiers' being "Sick / For breathing and exploit" (I.ii.16–17).[18]

Accordingly, the fool plays an important role, especially in providing mirthful companionship for the old Countess of Rossillion who, because of her age, would be more susceptible to the coldness and dryness of melancholy and would be all the more likely to benefit from Lavatch's presence.[19] Nor do the clown's efforts go unnoticed: the elderly courtier Lafew, who upbraids Helena's excessive mourning, is willing to indulge Lavatch (IV.v.68). Likewise, the Countess seems aware of the benefits of retaining her late husband's favorite fool, admitting that time spent listening to Lavatch's foolery is time well-spent: "I play the noble housewife with the time / To entertain't so merrily with a fool" (II.ii.62–63). On one level, then, Lavatch's fooling checks the effects of his mistress's aging by opening the channels from her heart and by warming and thinning her blood, thus prolonging her life. On another level, his humor and bawdry work against the elegiac mood of the play itself; his treatment of the Countess parallels Helena's healing of the king—curing the body politic as well as the king's body—in addition to her relief from her own love-longing and Bertram's recovery from his corrupt snobbery.

As You Like It offers a more extended version of the fool as physician to both body and body politic. Like Lavatch, Touchstone makes his appearance in a disordered and "diseased" socio-political milieu;

he enters the play at the very moment Celia suggests merriment as a remedy for Rosalind, who is lamenting her imminent banishment and her unfulfilled love for Orlando. The incorrigible melancholic Jaques also finds some respite from his condition in the company of a fool: "When I did hear / The motley fool thus moral on the time, / My lungs began to crow like chanticleer" (II.v.29–31). After this dialogue, Jaques begins to recognize the immense value of fools and foolery, expressing their value as moral guides in decidedly medicinal language:

> *Invest me in motley; give me leave*
> *To speak my mind, and I will through and through*
> *Cleanse the foul body of th' infected world,*
> *If they patiently receive my medicine.* (II.v.58–61)[20]

This passage demonstrates acknowledgement, by a self-proclaimed melancholic, of the medicinal value of fools and foolery. Jacques, however, fails to discard his affected (and dearly held) melancholy; it is for Rosalind to remind him of the virtue of foolery: "I had rather have a fool to make me merry than experience to make me sad" (IV.i.27–28). Rosalind here directly opposes the fool to the melancholic—one wrapped "in a most humorous sadness" (19–20)—suggesting that one of the means of avoiding melancholy was to employ the services of a professional fool.

In *Twelfth Night*, we see a similar opposition of fool and melancholic. Both Olivia and Orsino appear to be suffering from melancholy induced by grief. The former, mourning for her brother, is identified as being "addicted to a melancholy" (II.v.202), and the Duke, pining for the unreceptive Olivia, displays the characteristics of a sufferer from melancholy: he expresses a love of solitude—"for I myself am best / When in least company" (I.iv.38–39)—and music. He also displays knowledge of the physiological effects of his love-melancholy:

> *There is no woman's sides*
> *Can bide the beating of so strong a passion*
> *As love doth give my heart; no woman's heart*
> *So big, to hold so much; they lack retention.*
> *Alas, their love may be called appetite,*
> *No motion of the liver, but the palate,*
> *That suffer surfeit, cloyment, and revolt.* (II.iv.93–99)

Positioning themselves as melancholics, both Olivia and Orsino are eager for Feste's presence, indicating the importance of fools as therapy for melancholy. The Countess threatens him with expulsion for his excessive absence, while Orsino seems desirous of Feste's company, requesting the fool's singing: "That old and antique song we heard last night; / Methought it did relieve my passion much" (II.iv.4–5). In both instances, Feste provides some sort of remedy. He performs for Orsino and treats Olivia in a manner resembling Will Somers's series of questions, directly ministering to the source of her grief: "The more fool, madonna, to mourn for his soul being in Heaven" (I.v.76–77). Olivia, after the fool "mends," begins to display a healthy taste for mirth, joking with him about Sir Toby's drunkenness. Relieved of her sorrow, she is ready to begin a love relationship with "Cesario." Of course, both nobles (as Feste realizes) are to an extent affecting their passions, assuming the popular pose of the "witty" melancholic. Olivia makes this posturing evident in her rapid abandonment of her mourning in order to court "Cesario"; Orsino equally abruptly shifts his affection from Olivia to Viola. In both cases, Feste's cavalier attitude towards his employer may well indicate that he recognizes her sorrowfulness to be primarily a pose, as his ironic prayer, on Orsino's behalf, to "the melancholy god" (II.iv.73) suggests. It is, in one sense, to the foolishness of melancholic indulgence itself that Feste administers.

In another sense, however, such affectations of pseudo-Aristotelian melancholy may develop into a more serious malady. Viola's fictitious sister, pining away and (perhaps) dying from a "green and yellow" melancholy (II.iv.113), indicates the possibility that one assuming a melancholic pose may, in Chaucer's words, be in danger of making "ernest of game." Such a problem might be even more acute in Olivia, who, as Feste correctly observes, is squandering her childbearing years on excessive mourning, a passion which would in fact shorten her youth, not to mention her suitability for conceiving and rearing children.[21]

In contrast to *The Comedy of Errors*, where disease and madness remain purely illusory, the reality of death and disease underscore the comic "dream-world" of *Twelfth Night*: Olivia's brother is literally dead, and the darker side of statecraft has resulted in the maiming of Orsino's nephew; it nearly results in the death of Antonio. Even Sir Toby and Sir Andrew are forced, by the play's end, to suffer real wounds. In this more sophisticated play, therefore, with its tragic undercurrents, the fool's healing power becomes emblematic of the "patch" that comedy provides for tragedy. Feste alludes to the function of "patching" and "mending" in his first speech to Olivia:

> *Two faults, madonna, that drink and good counsel will amend, for give the dry fool drink, then is the fool not dry; bid the dishonest man mend, he is no longer dishonest; if he cannot, let the botcher mend him. Any thing that's mended is but patch'd; virtue that transgresses is but patch'd with sin, and sin that amends is but patch'd with virtue.* (I.v.42–49)

Appearing in the first act, this passage serves as an indication of the clown's function as a healer in the comedy. The "patching" and "mending," carrying obvious medical connotations, reflect the power of mirth and comedy to provide a respite from the reality of death, cruelty, and madness. Olivia's line "doth he not mend?" (I.v.73–74) invites a double reading: as Feste's improving of his own fooling and as his "mending" of the potential tragedy underscoring the play. Feste's fooling, indicative of the comic mood itself, soothes—or at least smooths over—sorrow and melancholic afflictions, a service he conveys to the audience in his concluding line "we'll strive to please you every day" (V.i.408).

Another way in which Shakespeare dramatizes the fool's healing power is by verbally allying this character with a real healer. Such an association occurs when Feste assumes the persona of Sir Topas the curate to "heal" Malvolio.[22] In the guise of Sir Topas, Feste recommends the healthful virtues of sleep: "Endeavor thyself to sleep, and leave thy vain bibble babble" (IV.ii.96–97). Further, the proverbial quality of the line "The fool shall look to the madman" (I.v.136–37) at least suggests general acknowledgement of the fool's ability to treat madness, an affliction usually associated with melancholy. This verbal alliance, in fact, predicts the more complicated relationship between healer and fool evident in *King Lear.*

This tragedy, presenting perhaps Shakespeare's most complex deployment of foolery as medicine, prevalently features medical discourse. Goneril refers to Lear's old age as his "infirm and choleric years" (I.i.293–94). The king's humoral condition—which makes him highly susceptible to adust melancholy—both reflects and contributes to his mental and intellectual shortcomings, as Regan adds to Goneril's physiological assessment: "He [Lear] hath ever but slenderly known himself" (I.i.293-4). Lear likewise employs physiological discourse in reference to his daughter's moral corruption, addressing Goneril as a corruption in his own bloodstream: "Thou art a bile, / A plague-sore, or embossed carbuncle, / In my corrupted blood" (II.iv.223–25). He later seeks a physiological cause for Regan's cruelty: "Then let them anatomize Regan; see what breeds about her heart. Is there any cause in nature that makes these hard hearts?" (III.vi.76–78).

In this medicinal framework, Lear's infirmity may be to a large extent attributed to melancholy. The "rising" of the heart he experiences, as well as the abdominal vapors, are both characteristic of excessive melancholy, which causes pernicious vapors to rise from the stomach, spleen, and intestines to the brain, in turn engendering insanity.[23] With his judgment already impaired, most likely by a form of melancholy common to the aged, his rage at his daughters could heat the mixture of natural melancholy and choler in his blood, engendering melancholy adust and madness. Bright, describing this form of melancholy, provides perhaps the most applicable case study for Lear: "If choller have yeelded to this sharpe kind of melancholie [melancholy adust], then rage, revenge, and furie, possesse both hart and head, and the whole bodie is carried with that *storme*, contrarie to persuasion to reason: which hath no farther power over these affections, then by way of counsell to give other direction."[24] Further, the ingratitude of Lear's children towards him, the initial cause of the king's collapse, has been cited as a factor inducing melancholy and madness. Elyot, who identifies among the effects of sorrow a weakening of the judgment and possibly even death, devotes an entire chapter to the ill effects of this feeling: "This affection [ingratitude] nyppeth the harte, ye of moste wyse men, for they love moste hartily, not provoked by carnal affection, but rather by good opinion."[25] In medicinal terms, therefore, Lear clearly embodies the interrelationship of a sorrowful mind and a blasted body ruined by melancholy. Already humorally imbalanced and prone to melancholy, the ingratitude Lear suffers could only exacerbate his condition.

The diseased body of the king, further, becomes a microcosm for the macrocosm of the divided realm; predictably, medicinal references underscore the downfall of the king and the kingdom. Lear's inability to maintain political authority translates into his vain attempt to regulate his own bodily humors. Aware of his own mental and physical deterioration, he expresses a fear of rising passions: "*Hysterica passio,* down, thou climbing sorrow, / Thy element's below" (II.iii.56–57), and "O me, my heart! My rising heart! But / down!" (121–22). Lear's commands to his body herald both the stripping away of his personal authority—he loses control first of his government, then of his family, then of his body, and finally of his mind—and the collapse of the kingdom's political cohesiveness.[26] In this way, Shakespeare fully develops an intricate relationship between mind and body, and between individual body and body politic, predicted in earlier plays.[27]

As in the earlier plays, the Fool plays an important role in this paradigm of body, mind, and state. It is true that, in a tragedy as bleak as *Lear,* the prospect of mirth as a possible remedy for Lear's melancholy-induced madness seems improbable; yet, when Lear feels madness encroaching upon him, it is to the Fool he cries first: "O Fool, I shall go mad!" (II.iv.286). It appears that Lear expects some sort of aid from his "all-licensed" court jester; indeed, the Fool does at least attempt to treat his lord's madness. Those loyal to Lear also suggest a medicinal function to the Fool's jesting: during the storm, Kent observes that Lear is accompanied by "None but the Fool, who labors to outjest / His heart-strook injuries" (III.i.16–17). It may be objected, of course, that the Fool's dialogues with Lear seem at times more chiding than jesting. However, chiding itself is an occasionally recommended part of the treatment for melancholy, as would be the mere act of providing companionship for the sufferer, both of which the Fool does.[28] Significantly, it is not until the Fool has left the play permanently (after Act III: "And I'll go to bed at noon") that Lear is seen as truly mad. Perhaps the Fool's absence itself is significant, marking the inevitability of a tragic outcome: the protagonist has descended beyond recall into melancholy or despair. Indeed, the only other play depicting the death of a fool, *Hamlet,* likewise concerns the impending insanity of the hero.

An understanding of the fool's role in treating melancholia may shed some light on the curious lexical and thematic links between Lear's Fool and Cordelia. This association, having inspired a long-standing stage tradition of casting the same person in the roles of the Fool and Cordelia, becomes explicit with Lear's repeated lines "Nothing shall come of nothing"—spoken first to Cordelia and then to the Fool—and culminates with Lear's carrying Cordelia's corpse while crying,

"And my poor fool is hang'd" (306).[29] In fact, Cordelia, whose name suggests the heart, is associated with both the Fool and physicians throughout most of the drama. She is allied more than once with physician figures. For instance, Kent, referring to himself as Lear's physician—"Kill thy physician, / And the fee bestow upon the foul disease [Goneril and Regan's flattery]" (I.i.162–63)—creates a dramatic alliance of physician (Kent) and fool (Cordelia). Both attend the king, and both are instrumental in his (temporary) recovery; it is Cordelia who directs Lear's recovery after he arrives in France. Nor should it be seen as coincidental that Lear's second and fatal collapse occurs right after Cordelia's death, just as his first mental breakdown follows the disappearance of the Fool. In both instances, the death of the Fool signals the complete mental and physical collapse of Lear, and, ultimately, the apocalyptic collapse of social order: "Is this the promis'd end? / Or image of that horror?" (V.iii.263–64).

King Lear thus presents the canon's most complex instance of a fool acting as a physician and attempting to heal both body and body politic. Although I have by no means offered an exhaustive study of the medicinal functions of Shakespeare's fools, my study has attempted to show Shakespeare's awareness and dramatic manipulation of the close interrelationship between mental and physical suffering, and the health-restoring value of mirth, expounded upon by Elizabethan medical tracts. Finally, the therapeutic value of fools provides an integral part of Shakespeare's deployment of medical imagery and discourse, especially important to his notion of treatment for bodily ailments as reflecting the "healing" of an ailing political "body." Therefore, the fool's role should be considered as something more important than modern scholarship has thus far recognized. This character is not simply a hired comedian: he or she is a "physician" who specializes in the purging of melancholy and sorrowful passions, thereby maintaining the health of both ruler and realm.

Notes

1 The list of scholarly works on fools and foolery is much too extensive to be comprehensively dealt with here. Any study of the subject in Renaissance society should begin with Enid Welsford's seminal work *The Fool: His Social and Literary History* (London: Faber and Faber, 1932) ch. III (55–75); Welsford suggests, among other reasons, that nobles

maintained fools to satisfy their taste for the grotesque, whereas Robert Hillis Goldsmith, *Wise Fools in Shakespeare* (East Lansing: Michigan State UP, 1955), refers to the fools' value in checking aristocratic arrogance, commenting that the fool's duty was "to entertain their masters and mistresses and at the same time to minister to their sense of self-importance" (48). Anton C. Zijderveld, *Reality in a Looking Glass: Rationality through an Analysis of Traditional Folly* (London: Routledge and Kegan Paul, 1982) offers a contrary view; in contrast to both Welsford and Goldsmith, Zijderveld argues that kept fools actually underscored the absolutist authority of Renaissance rulers: "The real point was that the monarch, by allowing his fool this kind of impudence, demonstrated in a rather painful manner how utterly powerless the rest of the court actually was" (120). William Empson, *The Structure of Complex Words* (London: Chatto and Windus, 1951), taking a lexicographic approach, emphasizes the malicious connotations of the word "fool" as "to dupe, make a fool of" (134).

2 The tendency among critics has been to acknowledge foolery as restorative, but only in an abstract sense. Empson considers this notion, only to reject it: "But I am not sure that we need regard this [the Fool's attempts to 'outjest' Lear's 'heart-struck injuries'] as a rational kindness towards Lear; the ideas belong to magic more than to medicinal psychology" (135). However, Empson, like Spillane, seems to have overlooked Elizabethan medical psychology in his assessment of the value of jesting. For further discussion of the fool as preserver of life and vitality in a symbolic sense, see Roger Ellis, "The Fool in Shakespeare: A Study in Alienation," *Critical Quarterly* 10 (1968): 245–68, and Harry Kavros, "The Crucifixion of Lear's Fool," *Neuphilologische Mitteilungen* 82 (1981): 305–6.

3 Brian Spillane, "'To Cleanse the Body O'Th' Infected World': The Fool and Therapy," diss, U of Dallas, 1990, 158.

4 James Black, "Shakespeare's Mystery of Fooling," *Mirror Up to Shakespeare: Essays in Honour of G. R. Hibbard*, ed. J. C. Gray (Toronto: U of Toronto P, 1984) 98.

5 Hardin Craig, "Shakespeare's Depiction of Passions," *Philological Quarterly* 4 (1925): 300.

6 Although all four substances were present in the body, each person had a prevalent humor which dictated his or her personality type. For the widespread interest in the melancholic humor, we may consider the number of authors writing about melancholy. Most prominent were Timothy Bright, *A Treatise of Melancholie* (1586; New York: Columbia UP, 1940), and Robert Burton, *The Anatomy of Melancholy*, ed. Holbrook Jackson (1640; London: Dent, 1932). Burton deals at some length with the interrelationship of sorrow and melancholy: "they beget one another

and tread in a ring, for sorrow is both a cause and a symptom of the disease" (I.258).

7 Although the *OED* mentions the word "melancholy" in a strictly psychological context as early as 1592, it also records its use as a physiological condition in a 1732 medical tract: "All spices are bad for melancholy people" (574). There would be no reason to believe, therefore, that when Shakespeare refers to "melancholy" or to a person inclined to that humor, he means a strictly psychological condition.

8 William Vaughan, *Naturall and Artificial Directions for Health* (London, 1600) 51. See also Andre du Laurens' *Discourse on the Preservation of the Sight*, trans. Richard Surphlet, Shakespeare Association Facsimile 15 (1599; Oxford: Oxford UP, 1938), and Thomas Wright, *The Passions of the Minde in Generall* (1604; Urbana: U of Illinois P, 1971).

9 For an example, see Samuel Rowland, *Democritus, or Doctor Merry-man: His Medicines against Melancholy Humour* in *The Complete Works of Samuel Rowland* (1607; New York: Johnson Reprint Corporation, 1966). The poetic epistle to his jestbook describes the famous Salernian "doctor":

> *And men of wit will unto him resort:*
> *To drive away dull melancholy mind,*
> *Which to madding frenzy is inclyn'd.* (29–31)

The association of humor and the purging of melancholy continued beyond the Renaissance; a 1729 collection of humorous songs bore the title *Pills to Purge Melancholy.* Such a late reference indicates that the physiological aspects of melancholy (and the purgation thereof) was understood well into the eighteenth century.

10 Robert Armin, *A Nest of Ninnies* in *A Nest of Ninnies and Other Jestbooks of the Seventeenth Century*, ed. P. M. Zall (Lincoln: U of Nebraska P, 1970): 48.

11 Armin, 59. Will's final riddle—the one which takes effect—is his coarsest: "'Now tell me,' says Will, 'if you can, what is it that, being born without life, head, lip, or eye, yet doth run roaring through the world till it die?' 'This is a wonder,' quoth the King, 'No question, I know it not.' 'Why,' quoth Will, 'it is a fart.' At this the king laughed heartily and was exceeding merry" (60).

12 Welsford 132–33. Matello's remedy was so effective, in fact, that Alfonso sent for him to relieve another illness, declaring that "her [Isabella's] gift [of a fool] was more welcome to him than the gift of a beautiful castle would have been" (133).

13 Welsford 132.

14 Other recommended cures, to which Shakespeare also refers, include music, sport, and rest. For a discussion of the medicinal effects of music

in the plays, see F. D. Hoeniger, "Musical Cures of Melancholy and Mania in Shakespeare," *Mirror up to Shakespeare: Essays in Honour of G. R. Hibbard,* ed. J. C. Gray (Toronto: U of Toronto P, 1984): 82–98. For the value of food, sport, and rest, see George Walton Williams, "Shakespeare's Metaphors of Health: Food, Sport, and Life-Preserving Rest," *Journal of Medieval and Renaissance Studies* 14.2 (Fall 1984): 187–202.

15 For the remainder of my study, I will confine myself to Shakespeare's professional fools, normally witty servants hired exclusively for foolery, namely, Touchstone, Feste, Lavatch, and Lear's nameless Fool. I mention Dromio first because he provides evidence for the medicinal value of fools in one of Shakespeare's earliest plays, and second, because Antipholus lists foolery as an important part of Dromio's job and in fact affords him something resembling a fool's license: "Because that I familiarly sometimes / Do use you for my fool, and chat with you ...." (II.ii.26–27).

16 See the abbess's speech (V.i.68–86) cited above.

17 Ann Barton describes the play as "concerned to evoke the remembrance of better times ... an almost Chekhovian backwater, elegiac and autumnal, filled with nostalgia for the past." Quoted in "Critical Introduction to *All's Well That Ends Well.*"

18 Black argues for Lavatch's importance to the milieu: "Lavatch with his cheerful bawdry helps keep both good humor and life itself flickering in the Rossillion household" (85–86).

19 According to Du Laurens, as a person ages "the action or operation of our naturall heat, which consisting in the consuming of his radicall moisture, doth by little and little fall a drying and cooling of our bodies" (171–72).

20 Critics such as Timothy G. A. Nelson, who compares Shakespearean fools to clergymen, have noted that the fool relieves Jaques's humoral condition: "Jaques finds Touchstone's mock sermon on time and decay healthy, refreshing, and funny. It *purges his melancholy humour.*" "The Fool as Clergyman (and Vice-Versa): An Essay on Shakespearian Comedy" in *Jonson and Shakespeare,* ed. Ian Donaldson (Atlantic Highlands, NJ: Humanitas P, 1983): 8 (italics mine).

21 As Viola tells her: "Lady, you are the cruell'st she alive / If you will lead these graces to the grave" (II.i.241–42). For a description of the way sorrow and melancholy bring about premature aging, see Thomas Walkington, *The Optick Glasse of Humours* (London: 1638): "A melancholic man is leaden, lumpish, of an extreme cold and drie nature, which cuts short the threed of his life long before it is spun" (65). See also Cymbeline's threat to Imogen (cited above).

[22] The "self love" with which the steward is said to be literally "sick" (I.v.90) was recognized as a physiological as well as psychological condition. According to Wright, "self-love upstarts, and for the affinitie with the sense ... wil in no case obay reason, but allured with the baite of pleasure and sensualitie, proclaimeth warres and rebellion against prudence, against the love of GOD .... In summe, from this infected love, sprung all the evils, welnie, that pester the world" (13).

[23] Du Laurens identifies this as "windy" or "hypochondriac" melancholy, which "ariseth from among the bowels, but especially from the spleene, liver and the membrane called mesenterium" (88). Further, like Lear, who has his lucid moments, the sufferer from this form of melancholy was afflicted sporadically: "it hath its periods, oftentimes making truce with the diseased" (87). That Shakespeare was aware of the notion that vapors—either gross or beneficial—ascended from the body to afflict the brain is demonstrated when Falstaff explains the ascendancy of warm and dry vapors generated by sack:

> *A good sherris-sack hath a twofold operation in it. It ascends me into the brain, dries me there all the foolish and dull and crudy vapors that environ it, makes it apprehensive, quick, forgetive ... becomes excellent wit.* (*2 Henry IV* IV.iii.96–102)

[24] Bright 112 (emphasis mine).

[25] Elyot 65. Elyot had previously claimed, referring to *Ecclesiastes,* that "Sorrow hath kylled many, and in itselfe is founde no commoditie .... this is so puissant and ennemye to nature and bodily helth, that to resiste the malyce and violence thereof, are required remedies" (64). For the effects of sorrow on the heart, see Wright 61–64.

[26] The best illustration of this concept in the canon is Menenius's "petty Tale" of the body parts' rebellion against the belly: "The Senators of Rome are this good belly / And you the mutinous members" (*Coriolanus* I.i.148–9).

[27] See the *Henry IV* plays.

[28] Du Laurens advises that "Melancholic persons should never be left alone, they should have some such companie left with them as might best like and please them" (106) and that "some whiles they must be chid for their foolish imaginations" (107).

[29] Lear's reference to Cordelia as his "fool" has opened many interpretive possibilities. Empson sees Lear's confusion of their identities as evidence of his madness. For a discussion of the dramatic implications of casting the same person as Cordelia and the Fool, see Richard Abrams, "The Double Casting of Cordelia and Lear's Fool: A Theatrical View," *Texas Studies in Literature and Language* 27 (1985): 354–368.

From *Rosalynde* to *As You Like It*: Shakespeare's Celebration of Blood Order

by Dilin Liu and Anumarla Govindan

John Russell Brown and Peter Erickson suggest in their separate studies that Shakespeare in *As You Like It* is concerned with the ideal of order—with "love's order" in Brown's terms or with "patriarchal order" in Erickson's interpretation.[1] They each argue convincingly that the play begins with a chaos resulting from the disruption of the order in question and ends in a harmony achieved through a restoration of the conventional order. However, while the two critics have correctly addressed Shakespeare's concern with "love's order" or "the patriarchal order," they have overlooked another very important order—the blood (class) order—with which Shakespeare, we believe, is equally concerned. A close examination of the play in comparison with its major source—Thomas Lodge's romance titled *Rosalynde*—will reveal that Shakespeare departs significantly from his source to widen the distance between his two classes, armigerous and base, and to teach the absolute importance of maintaining the established blood order in society.[2]

Judy Z. Kronenfeld, to our knowledge, is the only critic who has discussed Shakespeare's interest in the blood order in the play.[3] Unfortunately, she misinterprets Shakespeare's intentions by reading egalitarian beliefs into it. She proposes that Shakespeare, by underscoring the disparity and tensions between the high and low classes, is suggesting that the ideal egalitarian relationships in the pastoral tradition are but an illusion and must be "restored—or—created before they can be celebrated."[4] This interpretation of Shakespeare as an advocate of egalitarianism is inaccurate. Contrary to what Kronenfeld and other egalitarian critics would like to believe, Shakespeare, as David Shelley Berkeley argues in his study, is an "arch-conservative" who most obdurately insists on "the merits of the gentry and the demerits of the baseborn."[5] The gentry in Shakespeare's plays, except the degenerate ones, are always more virtuous and more worthy than the baseborn, and the former are always in control of the latter. This is because, to Shakespeare as well as to many of his contemporaries, the ranking of human beings according to their birth is a natural, God-ordained hierarchy that must not be violated.[6] "When degree is shak'd, / Which is the ladder of all high designs, / The enterprise is sick," as Ulysses declares in *Troilus and Cressida* (I.iii.101-3).[7] The belief in the ideal of blood order is shown in many of Shakespeare's histories and tragedies, where chaos arises either because someone (often a degenerate gentle or a bastard such as Claudius in *Hamlet*[8]) usurps the power of the rightful ruler or because the rightful ruler is not the apotheosis of the country. It is only after the rightful ruler is revenged or assumes the supreme power that order is restored and life goes back to normal.

This theme—the violation and restoration of the established blood order—is, however, not only found in the history and tragedy plays, but it is also evident in *As You Like It.* In adapting Lodge's *Rosalynde* to this comedy, Shakespeare amplifies the merits of the non-degenerate gentry, diminishes the virtues of the base, highlights people's tendency to strive for higher class status, and advocates the importance and necessity of upholding the existing blood order. In the following discussion, we shall therefore offer, first, a one-by-one comparison of Shakespeare's major characters with their counterparts in Lodge's *Rosalynde,* then an examination of the struggles over class status in the play (most of which are Shakespeare's free inventions), and finally a look at the denouement that Shakespeare created for the play as a celebration of the restoration of the God-ordained blood order.

To see how Shakespeare highlights the blood theme and amplifies the merits of the gentry, let us first turn to Orlando, a key character in *As You Like It.* The play opens with his complaining about his brother's treating him as a baseborn. Although Rosader (Orlando) of *Rosalynde* also speaks to his brother, Saladyne (Oliver), about the matter, Orlando's reference to blood is much stronger and much more dramatic. All Rosader says to Saladyne is:

> *I am thine equall by nature, though not by birth, and though thou hast more Cardes in the bunch, I have as many triumphs in my hands as thyselfe. Let me question with thee, why thou hast seld my Woods, spoyled my Manner houses, and made havock of such utensals as my father bequeathed unto me? I tell thee Saladyne, either answer me, or I will trouble thee as an enemie.*[9]

In contrast, Orlando lays far greater emphasis on blood matters:

> *I know you are my elder brother, and in the gentle condition of blood, you should so know me. The courtesy of nations allows you my better, in that you are first-born: but the same tradition takes not away my blood, were there twenty brothers betwixt us. I have as much of my father in me as you, albeit I confess your coming before me is nearer to his reverence.* (I.i.43-45)

By repeating his understanding of Oliver as the firstborn, Orlando honors more conspicuously than Rosader the custom of primogeniture—part of the established blood order of the day.[10] In the speech, Orlando also asserts that being the younger brother does not take his gentle blood away, a statement (nonexistent in Rosader's speech) underscoring Oliver's infringement upon Orlando's God-ordained blood privileges. Shakespeare's alteration in this matter does not stop here. In Lodge, after Rosader's complaint, Saladyne calms his brother by lying to him: Saladyne says that he will return Rosader's due property, and the two are at once temporarily reconciled. In Shakespeare, Oliver, instead of appeasing Orlando, strikes the latter and, what is worse, calls him "villain"; Orlando flares up in response:

> *I am no villain; I am the younger son of Sir Rowland de Boys. He was my father, and he is thrice a villain that says such a father begot villains. Wert thou not my brother, I would not take this hand from thy throat till this other had pull'd out thy tongue for saying so.* (I.i.54-59)

Orlando's statement best manifests the blood-consciousness of the sixteenth-century gentleman. For a gentle, the worst offense was to be called a villain—a baseborn (the word *villain*, according to *OED*, then meant "a low-born base-minded rustic" as well as an evil man).[11] Orlando goes on to tell Oliver that "You have train'd me like a peasant, obscuring and hiding from me all gentleman-like qualities. The spirit of my father grows strong in me, and I will no longer endure it" (I.i.54-58). With all these alterations and additions here, Shakespeare highlights both the dichotomy of the two classes and Orlando's worthiness as a born gentle.

To further increase the merits of Orlando, Shakespeare has Oliver admit, "Yet he's [Orlando's] gentle, never school'd and yet learned, full of noble device ..." (I.i.157-58). With this compliment, which does not exist in Lodge, Shakespeare bestows upon Orlando innate knowledge, a peculiar trait often found in the gentle, especially the royal, characters of his other plays. For example, the two princes in *Cymbeline,* Guiderius and Arviragus, though raised in the remote countryside, possess almost all the gentle qualities, including an unusual "invisible instinct" which, as David Bevington argues in his introduction to the play, "prompts them to assert their royal blood."[12] The two princes' foster father Belarius's monologue testifies clearly to their royal instinct:

... though trained up thus meanly
I' th' cave wherein they bow, their thoughts do hit
The roofs of palaces, and nature prompts them
In simple and low things to prince it much
Beyond the trick of others .... (III.iii.82–86)

Besides crediting Orlando with congenital knowledge, Shakespeare has made many other changes to increase Orlando's worthiness and virtue as a born gentleman. In *Rosalynde* Rosader is persuaded by Saladyne to challenge the wrestler (1:18), whereas in *As You Like It* Orlando is said to have intended to challenge Charles out of his own will (I.i.118-20), an act that characterizes a gentleman. In addition, Rosader is twice depicted as being melancholic (1:18, 51), a description Shakespeare dropped, because melancholy is a humor that, according to sixteenth-century belief, would cause degeneration of blood.[13] Instead of showing melancholy, Shakespeare's Orlando is full of a humor which, as John W. Draper indicates, is "sanguine" and "appropriate to 'noblemen.'"[14] This good blood explains why Orlando endures better than Adam on their trudge to Arden, whereas Lodge's Rosader becomes so weak that Adam has to offer his blood to sustain him. Furthermore, in Lodge, Saladyne makes Rosader his prisoner in their fight and then shows him publicly before his friends. Shakespeare left out these scenes because they are actions that, on the one hand, downgrade Rosader as a gentleman and, on the other hand, are something, as Draper points out, that "family pride would hardly have allowed in fact."[15] Finally, Lodge's Rosader hesitates for quite a while before he rescues Saladyne, while Orlando, on the contrary, rushes to Oliver's help without a second thought.

Besides glorifying Orlando, Shakespeare amplifies the merits of the other non-degenerate gentles. Duke Senior provides a good example. In *Rosalynde* Gerismond, the banished king, is presented as a very abstract figure who has few scenes to show his nobility. Duke Senior, in contrast, manifests his wisdom, noble morality, and leadership in, among other things, his conflict with Jaques, a degenerate gentleman Shakespeare created as a foil both to enhance Duke Senior's nobility and to emphasize the difference between the nondegenerate and the degenerate gentry (a point we shall discuss in detail later). In addition, whereas Gerismond converses unexceptionably in prose, Duke Senior speaks invariably in verse. The ability to use verse is generally one of the signs of a gentle character in Shakespeare.[16] More important, as Berkeley points out, Duke Senior, with his unusual mastery of poetic language, "turns his predicament into words and transcends his situation by style: Arden becomes 'a landscape of the mind,' a component of one's culture."[17] Shakespeare's efforts have turned the Duke, in Ralph Berry's words, into "an equable, urbane, and ideal philosopher ... and a man of rare mental equilibrium."[18]

Shakespeare also augments the merits of the two gentle female characters, Rosalind and Celia. He stresses Rosalind's eloquence and wisdom by giving her, as Edward I. Berry demonstrates, a far more prominent role in the play than Lodge does Rosalynde.[19] In addition, Shakespeare invents Touchstone and Jaques, who, again to use Berry's words, "serve as foils to Rosalind, setting off her distinguished qualities."[20] Celia, too, is elevated as a devoted friend to Rosalind, for she voluntarily accompanies Rosalind into exile whereas her counterpart, Alinda, is banished together with Rosalynde by her father. Moreover, Shakespeare extols the good manners of the court. Touchstone tells Corin that "If thou never wast at court, thou never saw'st good manners; if thou never saw'st good manners, then thy manners must be wicked" (III.ii.38-40). Although Duke Senior and some other characters have criticized the courtly life, their criticisms, we would argue, are more against Frederick's court than against courtly life in general. Otherwise, it is difficult to explain why Duke Senior is so ready to return to the court at the end of the play. Besides, when responding to Orlando's comment on manners, Duke Senior proclaims, "True is it that we have seen better days, / And have with holy bell been knolled to church ..." (II.vii.119-20). By "better days," he is, we believe, referring to the days under his own reign, and his statement is his indirect indictment of the corruptions under Frederick.

In his overall effort to glorify the gentry, Shakespeare even makes Frederick a less evil man who at least knows the gentry's code. When Oliver tells him that he (Oliver) has never loved his own brother, Frederick says to him, "More villain thou" (III.i.15). Although Lodge's Torismond also calls Saladyne "Nay villain," Shakespeare added the modifier "more" to stress the blood-concerned word *villain.* Moreover, Torismond calls Saladyne "villain" not to condemn the latter for hating his brother, but to blame him for causing the loss of Rosader: he tells Saladyne, "by thy meanes have I lost a most brave and resolute Chevalier" (1:60). Frederick's accusation, in contrast, focuses on Oliver's breaking of the gentry's code. According to Kelso, the most important virtues of the gentry in the sixteenth century included courtesy and friendship,[21] to which two virtues Edmund Spenser devoted two books (IV and VI) of *The Faerie Queene.*[22] At that time, courtesy "consisted of two parts, knowledge of what behavior was fitting to each man including oneself, and the graciousness in bestowing upon each his due."[23] In other words, a gentleman was required to be kind to others, including, obviously, his own family. Though it is ironic for the usurping Duke to denounce Oliver when he himself is behaving ungraciously to his own brother, a reader would not miss his message: unlike the base, the gentry love their family members. In fact, Frederick does follow this gentle code once when, as we have already mentioned, he banishes Rosalind alone, in contrast with Lodge's Torismond who banishes both his own daughter and Rosalynde. More important, Frederick repents at the end, manifesting a characteristic of the gentry (the base in Shakespeare's plays never repent), whereas Torismond sins to his very end—being killed in his attack on Gerismond's Arden.

To safeguard the honor and dignity of the gentry, Shakespeare also deliberately left out many scenes from his source, *Rosalynde.* For example, he divests Orlando of the violent actions that so often characterize Rosader, especially in the early stages of the novel. Rosader's rampage in Saladyne's house in an attempt to claim as his own Saladyne's estate finds no parallel in Orlando's actions. Orlando is too gentle to indulge in such foolish activities, especially activities in regard to the custom of primogeniture, as we have already demonstrated. In addition, Lodge vividly describes the fighting between Rosader and the Norman with gory details of the scene, leaving at least three dead in the presence of the ladies. Shakespeare, on the other hand, relegates the incident to a mere report far removed from the scene of action. In so doing, he avoids the embarrassment of depicting his gentle hero's stooping to fight with a baseborn churl, even though it is quite natural for gentles to fight with baseborns to put the latter in their place. By not assigning Oliver the gratuitous violence of Rosader, Shakespeare shuns lowering the dignity of a gentleman in the eyes of his audience.

While amplifying the merits of the gentry, Shakespeare diminishes the virtues of the base. Take Adam, for example. In *Rosalynde,* Adam offers his life (his blood) to save Rosader when they are being starved to death on their way to exile:

> *I am old, and overworne with age, you are young, and are the hope of many honours: let me then die, I will presently cut my veynes, & Master with the warme blood relieve your fainting spirits; sucke on that till I ende, and you be comforted.* (1:56)

Shakespeare deleted this episode from his play because he would neither allow a baseborn to make such a noble offer nor tolerate the idea of having a gentle drink a baseborn's blood. Such a deletion strips Adam of a virtuous action that Shakespeare would reserve only for the gentry. Apart from Adam, Corin and Silvius are also made less virtuous. In Lodge, Coridon (Corin) and Montanus (Silvius) provide Rosalynde and Alinda with the assistance they requested on first arriving in Arden: after Alinda explains their misfortune and makes requests for food and lodging, Coridon expresses both his deep sympathy and his willingness to help. Even though he acknowledges he is not able to relieve them totally, he tells the two ladies, "Marry, if you want lodging, if you vouch to shrowd your selves in a shepheards cotage, my house (for this night) shalbe your harbour" (1:46). Coridon's offer contrasts with Corin's behavior in *As You Like It,* wherein he declines Rosalind's request by insisting that he is in no position to help them because he is only "a shepherd to another man" who "is of churlish disposition" (II.iv.74, 76).

In addition to decreasing the virtues of the base characters he borrowed from Lodge, Shakespeare creates Touchstone, William, Audrey, and other low-borns to emphasize the demerits of the base. All these additional baseborn characters speak invariably in prose, as we have already pointed out, and (except for Touchstone, who has acquired much learning in the court) they are all pitifully ignorant. For example, Audrey does not even know the word *poetical*: she asks, "Is it [*poetical*] a true thing?" (III.iii.15-16). She cannot distinguish an adjective from a noun. William is overtly ridiculed for his ignorance by Touchstone (V.i.39–56). To stay consistent with his emphasis on the demerits of the baseborn, Shakespeare also deleted from his play Coridon's praise of the base over the gentry:

> *Louve (I tell thee) is as precious in a shepherds eye as in the lookes of a King, and we country swaynes intertain fancies with as great delight, as the proudest courtier doth affection .... and loyaltie ... is found more among shepherds than higher degrees.* (1:47-48)

As a believer in the blood order, Shakespeare would permit neither an egalitarian analogy between the King and a shepherd nor the propaganda of the baseborn's possessing more virtues than the gentry.

When the virtues of the gentry and the demerits of the base are made so obvious, maintaining the established blood order (the gentry over the base) appears very sensible. Furthermore, as it is human nature to strive for higher position in life, keeping the established order seems absolutely necessary. History proves the point: Kelso discovers in her study that in sixteenth-century England, "those who lacked the title [gentleman] were busy trying to acquire it, and those who had it were anxious to resist encroachment ...."[24] Shakespeare's play aptly reflects people's urge to rise in social rank. As Ralph Berry cogently argues, central to *As You Like It* is a drive for dominance, and the play consists of many "a struggle for mastery between two human beings."[25] Berry is entirely right in bringing to our attention all these struggles, but his interpretation of these struggles as being merely between "the romantic and the anti-romantic," or between the idealists and the realists, is questionable. In fact, the majority of the conflicts are, in the final analysis, over blood status. They are struggles either between the non-degenerate gentry and the degenerate ones, between the gentry and the base, or even between the base themselves for marginal superiority.

The two most important struggles in the play (Duke Senior versus Frederick, Orlando versus Oliver) concern blood status, and Shakespeare's alterations make the blood theme even more obvious. In *Rosalynde*, Gerismond and Torismond are not brothers. By making the two into siblings, and especially by making the banished Duke the senior, Shakespeare turns Frederick's usurpation of Duke Senior's position into a more apparent violation of the blood order—an atrocious breaking of primogeniture, a custom protected both by God and the law. In the fight between the other two brothers, Shakespeare, as shown earlier, has Orlando emphasize far more than Rosader his good blood, and thus he greatly intensifies Oliver's infringement of his brother's God-ordained class privileges. As is always the case in Shakespeare, chaos ensues with disruptions of the blood order. Frederick's crime leads to his daughter's leaving him and to his own mental uneasiness, for he is, as Le Beau tells us, "humorous" (I.ii.256). Oliver's infringement causes not only Adam's departure with Orlando but also Oliver's own downfall—Frederick seizes all of Oliver's property. The two degenerates' attempt to break the divine order thus meets with no success: Frederick returns his usurped position to Duke Senior after his repentance and conversion by "an old religious man" (V.vi.159-60), and Oliver ends up at the mercy of Orlando, who saves and, in turn, converts him.

Many other struggles occur in the play, all of which are Shakespeare's creations (since these are all waged by the characters he invented). First, we have Duke Senior's thwarting of Jaques's secret and ineffective challenge to the established order of life in Arden.[26] Jaques seems to have suffered from the most severe melancholy, "melancholy adust," which, according to Timothy Bright, is hot and dry, and wastes both the body and the mind.[27] Jaques shows all the symptoms of this form of melancholy, as Bright describes it: "the hart is made heavie, the spirits dulled, the cheerfull countenance altered into mourning, and life itselfe, which the nature of all things most desireth, made tedious unto persons thus afflicted."[28] A person with this kind of melancholy, as Berkeley explains, can never fully recover from it.[29] That is why Jaques is claimed to "be transformed into a beast" (II.vii.1). As so degenerate a person, he sinks deep into self-indulgence and complacency, which accounts for a fact that Helen Gardner points out: "the tyrannical Duke Frederick and the cruel Oliver can be converted; but not Jaques."[30] Because of his degenerate state, Jaques ignorantly questions Duke Senior and his followers' way of life (their killing of animals for food in particular), accusing the Duke of doing "more [to] usurp than doth your brother that hath banish'd you" (II.i.27-28). Jaques even naively considers himself an equal to Duke Senior: he boasts to Amiens, "I think of as many matters as he [Duke Senior]; but I give heaven thanks, and make no boast of them" (II.v.32-34).

Being a degenerate—an inferior in reality—Jaques is, however, simultaneously afraid to face the Duke, who possesses superior wisdom and morality. He admits to Amiens that "I have been all this day to avoid him [Duke Senior]. He is too disputable for my company" (II.v.31-32). Yet Jaques cannot escape his fate as a degenerate—being completely put down by the Duke. When Jaques brags about his wish to be a critic of the world, Duke Senior condemns him, driving him into his true position as a degenerate gentle:

> *Most mischievous foul sin, in chiding sin.*
> *For thou thyself hast been a libertine,*
> *As sensual as the brutish sting itself;*
> *And all th' embossed sores and headed evils,*
> *That thou with license of free foot hast caught,*
> *Wouldst thou disgorge into the general world.* (II.vii.64-69)

With the above chastisement, Duke Senior completely exposes Jaques' degeneracy and its cause—promiscuity, a trait that contradicts a gentleman's code. According to Lawrence Humphrey, a sixteenth-century gentleman should "thinke modestelye of him selfe, live temporatlye, and continentlye, behave hym selfe moderately, and soberly in all things."[31] Jaques possesses none of these gentleman's traits at all, a fact showing his extremely degenerate state. Hence, he also loses to Orlando (III.ii.) and Rosalind (IV.i.) in his strife for dominance over them. He is outwitted and humiliated by the two good members of the gentry. The only comfort Jaques has left as a degenerate gentle is flattering himself with the superiority of his birth over that of the base. In his exchange with Touchstone, Jaques, though enjoying no edge over the "fool" in knowledge or wisdom, may still belittle him. He declares that Touchstone's knowledge is "knowledge ill-inhabited, worse than Jove in a thatch'd house" (III.ii.8-9), and for all his knowledge, he is "yet a fool" (V.iv.103-4). Jaques's attitude toward Touchstone clearly reflects Shakespeare's insistence on the blood order—the dominance of the gentry, degenerate or not, over the base.[32]

A more interesting strife over class status is found in II. iv. where Touchstone, accompanying Rosalind and Celia to Arden, meets Corin and Silvius. Touchstone first calls Corin "clown" in an attempt to claim a superiority in rank, but Rosalind immediately retorts, "Peace, fool; he's not thy kinsman" (II.iv.62-63). Her response, as Kronenfeld observes, has a double meaning: it "puns on 'clown,' meaning court fool," and "it also inevitably applies the meaning Touchstone had intended for Corin to himself [a baseborn] ...."[33] Despite Rosalind's intention to put Touchstone into his class position, Touchstone as a court clown would not give up his marginal superiority over Corin as a shepherd. He answers Corin's "Who calls" with "Your better, Sir." Touchstone's effort to claim this marginal superiority continues throughout the rest of the play. He lectures William condescendingly, showing off his greater knowledge and better worth. As a result of Touchstone's insistence on his superiority, Corin calls him "master" (III.ii.11), and Audrey, William, and the first page all address him as "sir" or "gentleman," titles usually reserved for the gentry. Because of his marginal superiority, Touchstone defeats William in their strife for Audrey. But Touchstone is clever enough never to reach beyond his blood rank, for he understands that, though coming from the court, he is "yet a fool." He never competes with the gentry, except for the degenerate Jaques.

Touchstone is not the only baseborn character aware of his class. Corin, for example, understands his status, too: he refuses Rosalind's request for help because he claims, "I am a shepherd to another man / And do not shear the fleeces that I graze" (II.iv.74–75). Even Jaques, through his confrontation with Duke Senior and other gentle characters, finally learns of his own status. He realizes that, being degenerate, he is not entitled to be restored to his former rank. Hence, at the end of the play, he leaves the restored Duke and stays with Frederick in Arden.

The ending of the play is where Shakespeare departs furthest from Lodge in stressing his blood theme. *Rosalynde* concludes with an epilogue-like address, wherein Lodge announces that the moral of the romance is, among other things, "vertue is not measured by birth but by action ..." (1:139). The denouement supports such a theme: Gerismond rewards gentry and baseborns alike who have demonstrated virtues in their actions. Besides making Rosader his heir, Gerismond gentles the virtuous base, making "Montanus Lord over all the Forrest of Arden, Adam Spencer Captaine of the Kings Gard, and Coridon Master of Alindas flocks" (1:139). Shakespeare's message at the end of *As You Like It* is entirely different: he gentles no base;[34] rather, he ends his play by suggesting that worthiness is measured by birth. After restoration to his formal title, Duke Senior rewards his followers according to their rank rather than their virtue:

> *... every of this happy number*
> *That have endur'd shrewed days and nights with us*
> *Shall share the good of our returned fortune,*
> *According to the measures of their states.* (V. iv.171-174)

The word *states,* as both David Bevington and G. Blakemore Evans, two highly regarded editors of Shakespeare, define in their glosses, means "ranks."[35] Shakespeare's play is, therefore, a celebration of the restoration of the blood order disrupted by Frederick's usurpation of Duke Senior's position and also by Oliver's denial of Orlando's entitled privileges as a gentleman by birth.

Shakespeare made other alterations at the end of the play to highlight his blood theme. In both the source and Shakespeare's play, there is a wedding ceremony. However, in Lodge, the ceremony is disrupted by the usurping King's attack. All the men leave the scene for the battle. To the contrary, the wedding celebration in *As You Like It* is coupled with Duke Senior's restoration, giving a sense of double happiness. Moreover, whereas Lodge's romance shows a wedding of three couples, Shakespeare's play presents four (with that of Touchstone and Audrey added). The addition not only creates a sense of balance that the source lacks but also heightens the implication that there should be no crossing between the gentry and the base because all four marriages are exclusively between characters of matched class status, gentle-to-gentle and base-to-base. Therefore, coinciding with the restoration of Duke Senior and the other non-degenerate gentry, the celebration of these well-matched marriages becomes symbolically a celebration of the God-ordained blood order—an order providing the foundation of Shakespearean society.

Notes

[1] John Russell Brown, "Love's Order and Judgment of *As You Like It*," in *Twentieth Century Interpretations of* As You Like It, ed. J. L. Halio (Englewood Cliffs: Prentice-Hall, 1968) 70-87; Peter Erickson, "Sexual Politics and Social Structure in *As You Like It*," in William Shakespeare's *As You Like It*, ed. Harold Bloom (New York: Chelsea House, 1988) 113-130.

[2] Although quite a few scholars have conducted source studies of the play, no one so far seems to have related his or her findings to Shakespeare's blood theme in the play. See Geoffrey Bullough, *Narrative and Dramatic Sources of Shakespeare*, 8 vols (London: Routledge and Kegan Paul, 1958) 2:143-57; Kenneth Muir, *Shakespeare's Sources* (London: Methuen, 1978) 125-31; and Marco Mincoff, "What Shakespeare Did to *Rosalynde*," *Shakespeare Jahrbuch* 96 (1960): 78-89.

[3] Judy Z. Kronenfeld, "Social Rank and the Pastoral Ideals of *As You Like It*," *Shakespeare Quarterly* 29 (1978): 333-48.

[4] Kronenfeld 348.

[5] David Shelley Berkeley, *Blood Will Tell in Shakespeare's Plays* (Lubbock: Texas Tech Press, 1984) 7.

[6] The basic Renaissance concept of the class hierarchy, as Ruth Kelso in *The Doctrine of the English Gentleman in the Sixteenth Century* (Urbana: University of Illinois Press, 1929) 35 summarizes, is that "at the top was the ruler [the king], deriving his power directly from God, acting as God's viceroy upon earth. Under him ... must be a body of lesser authorities to assist him in the administration of justice. This body included dependent princes, magistrates, officials of all sorts, and the whole body of nobility, in whom was vested something of the divine authority of the king. At the bottom, furnishing support for the pyramid and the admitted reason for its existence, were the common people." Sir William Segar, *Honor Millitary and Civill* (London, 1602) 51, offers a more detailed classification of this hierarchy: "We in England doe divide our men into five sorts: Gentlemen, Citizens, Yeomen, Artificers, and Labourers. Of Gentlemen, the first and principle is the King, Prince, Dukes, Marquesses, Earls, Viscounts, and Barons. These are the Nobilitie, and be called Lords, or Noblemen. Next to these be the Knights, Esquires, and simple Gentlemen, which last number may be called nobilitas minor." See also Sir John Ferne, *The Blazon of Gentrie* (London, 1586) 85, on this matter.

7 All citations from Shakespeare's plays are from *The Complete Works of Shakespeare,* ed. David Bevington. 3rd ed. (Glenview: Scott, Foresman and Company, 1980).

8 For further information on the possibility of Claudius' being a baseborn, see David Shelley Berkeley, "Claudius the Villein King of Denmark," *Hamlet Studies* 11 (1989): 9–21.

9 Thomas Lodge, *Rosalynde: Euphues Golden Legacie* in *The Works of Thomas Lodge,* 4 vols. (1883; rpt. New York: Russell & Russell Inc., 1963) 1:8. Subsequent citations are from this edition. We have retained the original spelling.

10 Charles I. Elton, *Encyclopedia Britannica* (1895 ed.), points out that primogeniture was very popular in England during the Renaissance and that it was enforced "by law, custom or usage." According to *A Health to the Gentlemanly Profession of Servingmen* (1598; rpt. Roxberghe Library, 1868) 108-9, the reason for the custom was that dividing the family possessions among all the children would break the established high family.

11 "Villain," *Oxford English Dictionary,* 1989 ed.

12 David Bevington, ed., *The Complete Works of Shakespeare,* 3rd ed. (Glenview: Scott, Foresman and Company, 1980), 1416.

13 Timothy Bright, in *A Treatise of Melancholie* (1586, rpt. New York: Columbia University Press, 1940) 30, defines melancholy as a "humor no otherwise then that part of that bloud which naturally of the rest is most grosse; and the excrement the superfluitie of the same; which if it putrifieth, bestowth still the same of a farre diverse thing both in temper & nature called black choler. The melancholie passion is a doting of reason through vaine feare procured by fault of the melancholie humor." Petrus Pomarius Valentinus in *Enchiridion Medicum* (London, 1608) 16, claims that melancholy is "the dregs and sediment of bloud" that could become "vicious" and cause "madness." As for how melancholy causes degeneracy in blood in Shakespeare's characters, see Berkeley, *Blood Will Tell* 24-25, 56.

14 John W. Draper, "Shakespeare's Orlando Inamorato," *MLQ* 2 (1941): 180.

15 Draper, "Orlando, the Younger Brother," *PQ* 13 (1934): 75.

16 In Shakespeare's plays, the gentry usually speak in verse whereas the base almost always speak in prose. *As You Like It* is no exception. Duke Senior and Duke Frederick speak invariably in verse; the other gentles switch between verse and prose; and the base characters, except the

shepherds, converse always in prose. The shepherds Corin, Silvius, and Phebe are given exception probably because they are conventional pastoral figures who use verse, as the genre requires.

17 Berkeley, *Blood Will Tell* 22.

18 Ralph Berry, "No Exit from Arden," *Modern Language Review* 66 (1971): 13.

19 Edward I. Berry, "Rosalynde and Rosalind," *Shakespeare Quarterly* 31 (1980): 42-52.

20 Edward I. Berry 43.

21 See Kelso 70-85.

22 Edmund Spenser, *The Faerie Queene* (London, 1596).

23 Kelso 79.

24 Kelso 18.

25 Ralph Berry 11.

26 Though a few critics have suggested that Shakespeare creates Jaques as a critic to expose social evils, a close examination will reveal, as Robert B. Bennett, "The Reformer of a Malcontent: Jaques and the Meaning of *As You Like It*," *Shakespeare Studies* 9 (1976): 183, argues convincingly, that Shakespeare makes "him [Jaques] more an object of social criticism and Romantic correction than an authorial mouthpiece for anatomizing the ills of society." Jaques as an object of attack serves, as we will show, to illustrate Shakespeare's concept of blood order.

27 Bright 32-3.

28 Bright 33.

29 Berkeley, *Blood Will Tell* 24–5.

30 Helen Gardner, *As You Like It*, in *Twentieth Century Interpretations of* As You Like It, ed. Jay. L. Halio (Englewood Cliffs: Prentice-Hall, 1968) 67.

31 Lawrence Humphrey, *The Nobles or of Nobilitye* (London, 1563) riii.

32 Here we are not suggesting that the degenerate gentry cannot be worse than the base in the ethical sense; they can and often are, as Berkeley, *Blood Will Tell* 45-80, demonstrates clearly. But they cannot be lower than the base in rank.

33 Kronenfeld 344.

34 Gentling the base is not what Shakespeare would normally do. In his whole canon, there is only one instance of baseness being gentled—the mass gentling of the soldiers by the King in *Henry V* (IV.iii). Shakespeare's reluctance to gentle the base conforms to the popular belief of his

contemporary gentry. For example, John Selden, *Titles of Honor* (1614; rpt. London, 1631) 866, condemns the idea that gentlemen could be created: "And to say, that a gentleman in such a sense [blood sense] may be created, is itselfe a contradiction. No creation can make any man to be really of another bloud then hee is …."

35 David Bevington 391; G. Blakemore Evans, ed., *The Riverside Shakespeare* (Boston: Houghton Mifflin, 1974) 399.

"Thou Art a Villain": From the Ensign to Iago—Blood Changes in *Othello*

by Dennis F. Bormann

Only once in the canon does Shakespeare apply the term *villain* to a character in the *dramatis personae* of a play—that character is Iago. The *Oxford English Dictionary* defines *villain* as "originally, a low-born base-minded rustic; a man of ignoble ideas or instincts; in later use an unprincipled or depraved scoundrel; a man naturally disposed to base or criminal actions."

Though qualified as now in rare usage, the term's epistemology is further explained as meaning a person of "low or mean ... birth or position."[1] Most modern critics interpret Iago's villainy in the ethical sense, choosing to ignore the class meaning of the word, a meaning commonly understood in Shakespeare's day.[2] As John W. Draper relates in *The Humors and Shakespeare's Characters*, modern interpreters of the plays, in general, whether unconsciously or not, "endeavored, not so much to set forth the original authentic Shakespeare, as to read into him the ideas and the ideals of their own age."[3] Certainly Iago is a villain in the ethical sense—his actions speak for themselves—but an important, informing facet of the play is missed when the class sense of the word is unknown or ignored. In *Studies in Words*, C.S. Lewis feels that the term is never without "some implication of ignoble birth, coarse manners, and ignorance."[4]

This study will attempt to interpret Shakespeare's application of *villain* to Iago by examining the blood changes Shakespeare makes in adapting the source of *Othello, The Hecatommithi* by Giovanni

Baptista Geraldo Cinthio. As David Shelley Berkeley argues in *Blood Will Tell in Shakespeare's Plays,* throughout the canon, Shakespeare always intensifies "whatever class-consciousness may exist in their primary sources,"[5] and *Othello* is no exception; what is suggested by these changes in blood matters is that Iago is a villain in the class sense as well as in the ethical sense.[6]

Several factors support and develop this interpretation: Iago has no understanding of social graces; his language is gross and unattractive, unfitting for a gentleman; he is put down repeatedly in a class sense; he fails to stand up for himself when various members of the play directly attribute him with being base in origin, a slur no gentleman could endure; there is a lack of family solidarity—he and Emilia are not one spirit intertwined; like Shakespearean villains in general, he is without issue; he is passed over for the lieutenancy, a commission which would have bestowed on him the rank of gentleman; Iago will not issue an open challenge; he is totally selfish, espousing a philosophy of egotism; and lastly even when his evil deeds are out in the open, Iago never repents. My discussion will qualify and deal with some seeming incongruities the play presents with this argument: Why does Shakespeare allow Iago to speak in blank verse, something usually reserved for the gentry? And what of Iago's keen intelligence? There is also the important matter of Othello, who is of royal lineage, choosing Iago as his confidant, a matter that seems inconsistent given Shakespeare's class bias.

As previously noted, most critics choose to ignore the class-based division Shakespeare uses in his plays, applying instead their own modern theories of interpretation, but Berkeley goes further than Draper in feeling that many such critics are "imbued with the taboos of egalitarian liberalism" and avoid Shakespeare's class bias as "an unwelcome reminder of the *ancien regime*."[7] Certainly this is more true in some cases than in others, but a few contemporary critics do cite Iago's mean birth. Robert B. Heilman notes that Iago has many base traits, including his vocabulary, but Heilman feels these attributes apply as metaphors for Iago's baseness of spirit.[8] Marvin Rosenberg cites Iago's lower social class and the unmistakable identification of the Folio cast list.[9] In *Preface to* Othello, Harley Granville-Barker, after suggesting that Shakespeare molded a more heroic Othello out of Cinthio's Moor, says that Shakespeare also set up Iago "in total contrast to [Othello]; a common fellow, foulminded and coarse-tongued."[10] The critic whose sensibilities and arguments are closest to the focus of this essay is Draper who, in *The* Othello *of Shakespeare's Audience,* writes that Iago is a "low-born non-commissioned officer."[11] Draper bases much of his study on the "Elizabethan psychology expressed in the old theory of the humors derived from Galen,"[12] but, with a brief exception, he does not concern himself with the changes Shakespeare made from Cinthio's novel. This study will further develop, with more specific information, these alterations from *Othello*'s primary source, alterations that enhance the characterization and the plot of the play.

In 1565 Giovanni Baptista Geraldo Cinthio published *The Hecatommithi,* a book divided into ten decades, each decade devoted to a particular subject with ten corresponding stories. *Othello* is based on the seventh story in the third decade, its assigned subject being "The Unfaithfulness of Husbands and of Wives" (Cinthio 376).[13] No controversy exists over the validity of Cinthio's novel's being the source of *Othello* as there are other disputes over sources for some other plays of the Shakespearean canon, such as *The Taming of the Shrew,* nor is there any complication of multiple sources as there is in *The Comedy of Errors* and most of the histories.

In Cinthio's novel, the Ensign is characterized as

> *a man of handsome figure, but of the most depraved nature in the world. This man was in great favour with the Moor, who had not the slightest idea of his wickedness; for, despite the malice lurking in his heart, he cloaked with proud and valorous speech and with a specious presence the villainy of his soul with such art that he was to all outward show another Hector or Achilles.* (378)

This "valorous speech," with its comparison of the Ensign to noble soldiers of the Trojan war, implies the Ensign's understanding of social graces, as well as his mastery of a high quality of speech. In the source, the Ensign has no dialogue with Disdemona, but in *Othello* Shakespeare allows the audience to hear Iago's indelicate aside when Cassio and Desdemona speak intimately together:

> *He takes her by the palm. Ay, well said, whisper. With as little a web as this will I ensnare as great a fly as Cassio. Ay, smile upon her, do; I will gyve thee in thine own courtship. You say true; 'tis so, indeed. If such tricks as these strip you out of your lieutenantry, it had been better you had not kiss'd your*

> *three fingers so oft, which now again you are most apt to play the sir in. Very good; well kiss'd! An excellent courtesy! 'Tis so, indeed. Yet again your fingers to your lips? Would they were clyster-pipes for your sake!* (II.i.166–75)[14]

Iago thinks the kissing of a lady's hand a sexual pass, his vulgar mind perverting this gentle custom. It is quite possible that as a common soldier Iago is ignorant of this type of social intercourse and its implicit innocence—intercourse that is identified as proper in Count Baldassare Castiglione's *The Book of the Courtier,* translated into English by Sir T. Hoby in 1561.[15] Written in dialogue form, Castiglione describes such exchanges as "intercourse most free and honourable; for everyone was permitted to talk, sit, jest and laugh."[16] As Gerald M. Pincess and Roger Lockyer write of *The Courtier,* "Castiglione's speakers are concerned with social deportment. For them good and bad are matters of grace and style."[17] Iago does not realize that Cassio is only showing Desdemona the proper respect in his address and kiss.

The lack of understanding of social decorum also manifests itself in Iago's speech, for his language is gross; he constantly likens humans to animals, which again usually mirrors his salacious mind. Though identified as being "that wickedest of all bad men,"[18] the Ensign of Cinthio's prose work hides this depravity in "valorous speech." When he insinuates Disdemona's disloyalty to the Moor, the Ensign's dialogue exhibits a mannered style:

> *I can't deny it pains me to the soul to be thus forced to say what needs must be more hard to hear than any other grief; but since you will it so, and that the regard I owe your honour compels me to confess the truth, I will no longer refuse to satisfy your questions and my duty. Know, then, that for no other reason is your lady vexed to see the Captain in disfavour than the pleasure that she has in his company whenever he comes to your house, and all the more since she has taken an aversion to your blackness.* (381)

In contrast, Shakespeare presents Iago's speech as rough and vulgar, as when he shouts up to Brabantio's terraced position: "You'll have your daughter cover'd with a Barbary horse; you'll have your nephews neigh to you; you'll have coursers for cousins and gennets for germans" (I.i.111–14). Iago's pattern of equating humans to animals continues throughout the play, and the instances are too numerous to mention here. In the case just cited, he is directing these base allusions as racial slurs toward the possible grandchildren of Brabantio that could issue from Othello and Desdemona's miscegenation.

Shakespeare often disparages his base characters with this quality of speech, but in *Othello* this quality carries important implications for the plot. Ifor Evans sees this emphasis on Iago's quality of speech as "maintained with an almost intolerable insistence ... supported by a continuing reference to the mind and its emotions in terms of disease, or infection, or plagues of toads and flies."[19] Heilman explains that "in his barnyard view of life, Iago instinctively dehumanizes the human being, especially by treating love as a mechanical animality."[20] Draper writes that Shakespeare mirrors situation and character in his use of meter, Iago's being rough in contrast to Cassio's smooth-flowing lines.[21] Iago is always laying seeds of directed half-truths in his guttural speech, and to a modern reader this might be an argument for the power of suggestion, but the play suggests Shakespeare means it in a physical sense as well, a change registered in the blood. All things register in the blood, including thoughts and actions, "rendering this element the cause and talisman of what may be expected of human beings."[22] Iago's base mutterings to Roderigo are a prelude to the poison he injects Othello with, directing the crime against gentle Desdemona. Though Othello is unaware of his lieutenant's deceit, his blood is affected because "even involuntary participation in crime caused gentle blood to become somewhat gross."[23] Iago makes reference to this change himself when he states:

> *The Moor already changes with my poison.*
> *Dangerous conceits are, in their natures, poisons,*
> *Which at the first are scarce found to distaste,*
> *But with a little act upon the blood*
> *Burn like the mines of sulphur.* (III.iii.330–34)

Most modern critics do not interpret the change in the Moor as physical, but Shakespeare leaves us the strong implication that this is, in fact, the case. Though not seen, Iago's base mutterings cause a physiological change within Othello's blood. The only apparent change is in the Moor's behavior and his acceptance of Iago's vile accusations, but given Othello's noble blood this would not occur unless his blood had physically degenerated. This is backed up by Renaissance texts such as Robert Burton's *The Anatomy of Melancholy*, published in 1621. Burton writes, "So on the other side, the minde most effectually workes upon the Body, producing by his passions and perturbations, miraculous alterations."[24]

Throughout the play Iago is put down in a class sense by various characters. In Cinthio's version the Ensign is never directly confronted with any insinuation that he is of base blood. Even when the Moor is informed by the Ensign of the possibility of Disdemona's infidelity, the Captain, in his rage, makes no such claims against the man's social status, instead only issuing a threat to his officer: "By heavens, I scarce can hold this hand from plucking out that tongue of thine, so bold which dares to speak such slander of my wife" (Cinthio 381). During the exchange of dialogue between Iago, Desdemona, and Cassio in Act II of *Othello,* Desdemona replies to Iago's banter, "These are old fond paradoxes to make fools laugh i'th' alehouse" (II.i.137–38). As *Blood Will Tell* relates, ale is associated with the baseborn,[25] and the implication here is that Desdemona considers Iago base. This is further reinforced when Desdemona comments, "How say you, Cassio? Is he not a most profane and liberal counselor?" (II.i.162–63), to which Cassio replies, "He speaks home, madam" (II.i.164). Bevington glosses *home* as "without reserve," and Desdemona's use of *profane* and *liberal* point to the base trait of openness. Berkeley observes: "The epithet 'honest' as applied to a man … may be a pejorative expressive of upper-class disapproval of plebeian openness,"[26] and this certainly is the case here, as well as when Iago constantly is referred to as "honest Iago" by Othello.[27]

When Brabantio retorts to Iago's accusations in Act I, he calls him a "profane wretch" (I.i.115). Iago answers typically, using the aforementioned animal imagery, with his comment regarding "the beast with two backs" (I.i.117). Note that Iago makes reference to sex openly and in a vulgar manner, but just as important is that his non-denial of being a profane wretch signals that Iago is in fact base. Even when Brabantio directly calls him a villain—an insult a gentleman would not suffer— Iago never denies it.

Iago never stands up for himself, even when confronted with class slurs. Emilia, who does not know as yet of whom she speaks, ironically identifies the cause of Othello's change in temperament towards Desdemona with her tirade of "some eternal villain, / Some busy and insinuating rogue, / Some cogging, cozening slave, to get some office, / Have … devis'd this slander" (IV.ii.132–35). She follows with "The Moor's abus'd by some most villainous knave, / Some base notorious knave, some scurvy fellow" (IV.ii.141–42). Iago again does not deny it, but instead cautions her with "Speak within door" (IV.ii.146), meaning "not so loud." Iago does not care for his reputation. This is in direct contrast to Cassio, who laments the loss of his commission: "I have lost the immortal part of myself, and what remains is bestial" (II.iii.257–59). Iago replies, "Reputation is an idle and most false imposition" (II.ii.262). That Iago can make such a statement is telling in regard to the thesis of the present argument. As Henry Peacham wrote in *The Compleat Gentleman,* "There is no thing that setteth a fairer stampe upon Nobilitie then evenesse of carriage and care of our Reputation";[28] to be without reputation is to be "*dead long before we are buried.*"[29]

When Emilia unknowingly hits close to the mark with her accusations that some villain is abusing Othello, Iago tells her to keep quiet. She does not comply, but continues in complete disregard of being her husband's command; this defiance signals another aspect of Iago's plebeian rank—that of lacking any family solidarity: Iago and Emilia are not one spirit intertwined. John Draper notes that Emilia is of a lower social status than her counterpart in Cinthio's novel; in Shakespeare's source the Ensign's wife is found to be "a young, and fair, and virtuous lady" (Cinthio 378) on equal footing with Disdemona instead of her servant. The Ensign's wife also "knew the whole truth" about her husband's plans against her friend and the Moor, and though she does not condone such treachery, she does not "disclose a single circumstance" (Cinthio 384). Compare this maintained confidence with Emilia's outraged reaction at the end of the play. After finding out that Iago was the villain poisoning Othello's humor, she calls her husband a liar. Iago again, as before, commands her to be quiet and "charm your tongue" (V. ii.190). Emilia does not heed Iago, shouting:

> *Villainy, Villainy, Villainy!*
> *I think upon't — I think I smell it! O villainy!*
> *I thought so then — I'll kill myself for grief —*
> *O villainy, villainy!* (V.ii.197–200)

This defiance of her husband would have been understood to be in keeping with Emilia and Iago's social status as baseborns because Renaissance texts devised exacting prescriptions for marriage. In their study of English Renaissance texts, Pinciss and Lockyer discuss an official sermon sanctioned by the Church of England concerning marriage which reads: "This homily sets forth the received notion of the ideal relationship between husband and wife in marriage. Its position ... is particularly significant: husband's [sic] command and wives must obey."[30] There must also be an effort by both to "apply their minds in most earlies wise to concord"[31] so that "we may have the fellowship of our wifes."[32] Richard Brathwait in *The English Gentleman,* another Renaissance text of 1630, comments, "Wives are to submit themselves unto their Husbands; and again, how Husbands should love their Wives, Even as Christ loved the Church, and gave himselfe for it."[33] Marianne L. Novy, in her discussion of Renaissance love relationships, writes of mutuality in this way: "While the word itself [*obey*] does not necessarily imply equality or inequality, it implies sharing and companionship."[34] Henry Smith, a popular preacher in Elizabethan England, appropriated these similes about marriage to his *Preparation to Marriage*: "Husband and wife are like a pair of oars, a pair of gloves."[35]

Iago and Emilia have no such relationship. Heilman sees this lack of a solid partnership as a "measure for Iago" and cites Dante's doctrine that "the will is subject to the guidance of love, and that to will rightly one must love rightly."[36] Recalling that Iago's language is laden with bestial similes, Norman Sanders writes that for Iago, "spiritual values are debased," and "love is merely an anatomical function."[37] In the play, Iago says to Roderigo that love "is merely a lust of the blood, and a permission of the will" (I.iii.137). Theodore Spencer writes in *Shakespeare and the Nature of Man,* "it is obvious that [Iago] knows nothing about [love]."[38]

Evidently, no children are produced out of this dubious and base union between Iago and Emilia. Iago is without issue, like Shakespeare's villains in general, the only noted exceptions being Aaron in *Titus Andronicus* and possibly Lady Macbeth. One could speculate, with some degree of confidence, that this is due to a weakness in blood. Thomas Cogan, author of the Renaissance medical text *The Haven of Health,* describes semen as being transformed from the blood—and the better the quality of the blood, the more seed that is produced.[39] Cinthio's Ensign has "a little daughter, a child three years of age" (382). That Shakespeare deletes this offspring in his story is revealing, given Iago's villainy—it is in absolute keeping with Shakespeare's class-consciousness.

Another matter that informs one of Iago's plebeian status is Othello's passing him over for a commission in favor of Cassio. This element is also missing from the source of the play as a hint of impetus for the Ensign's treachery. In its stead is the complication of the Ensign's falling in love with Disdemona:

> *Now the wicked Ensign, regardless of the faith that he had pledged his wife, no less than of the friendship, fidelity, and obligation which he owed the Moor, fell passionately in love with Disdemona, and bent all his thoughts to achieve his conquest.* (Cinthio 379)

Shakespeare could not have this part of Cinthio's narrative applied to Iago. It is unthinkable for a baseborn in Shakespeare to aspire to love a gentlewoman. Certainly it would have meant Shakespeare's imbuing Iago with the virtuous emotion of love, something that as a baseborn he would have dubious claim to. This is one of the reasons Shakespeare invents Roderigo—to act as the spurned lover. But Shakespeare also invents that part of the play dealing with Iago's being passed over for the commissioned status which would have honored Iago with the rank of gentleman. Giovanni Nenna writes of this custom in *A Treatise of Nobility*: "Others make themselves and their posterity noble, in following the war, applying the same with all their indevour, albeit ... they doe descend but of base parentage."[40] This professional rebuff gives Iago motivation in performing his evil deeds, an important point missed by modern critics.[41] Iago confides to Roderigo at the play's opening that he hates the Moor because of this rebuff, saying:

> *Despise me if I do not. Three great ones of the city,*
> *In personal suit to make me his lieutenant,*
> *Off-capp'd to him; and, by the faith of man,*
> *I know my price, I am worth no worse a place.*
> *But he, as loving his own pride and purposes,*
> *Evades them with a bombast circumstance*
> *Horribly stuff'd with epithets of war....*
>
> [And, in conclusion,]
>
> *Nonsuits my mediators; for, "Certes," says he,*
> *"I have already chose my officer."*
> *And what was he?*
> *Forsooth, a great arithmetician,*
> *One Michael Cassio, a Florentine,*
> *A fellow almost damn'd in a fair wife,*
> *That never set a squadron in the field,*
> *Nor the division of a battle knows*

More than a spinster—unless the bookish theoric,
Wherein the toged consuls can propose
As masterly as he. Mere prattle without practice
Is all his soldiership. But he, sir, had th' election,
And I, of whom his eyes had seen the proof
At Rhodes, at Cyprus, and on other grounds
Christian and heathen, must be belee'd and calm'd
By debitor and creditor. This counter-caster,
He, in good time, must his lieutenant be,
And I—God bless the mark!—his Moorship's ancient. (*I.i.8–34*)

At first glance one might believe that Iago has a point. Why give the commission to Cassio, a scholar, rather than to Iago, a combat-tried soldier? The answer may lie with George Whetstone and his book of 1585, *The Honorable Reputation of a Soldier*. "It is not nough that the souldier be valiat in execution ... but obedience, is also especially required,"[42] and a "good souldier oughtte be studious in matters of Pollycie."[43] In other words there are many factors that probably went into Othello's decision, something Iago does not account for Iago's mind is enveloped in matter, sucked into immediacy. When he devises his "double knavery" (I.iii.395), it is in quick reaction rather than as a plan, showing his ineptitude for strategic thinking. Sanders calls Iago a "miscalculating improviser,"[44] an apt description given the results of his actions.

Perhaps this quality of opportunism governed Iago's military campaigns as well, leading him to seek advantage in situations rather than stride forth with courage and valor, because in the play Iago shows himself to be a coward; he will not issue an open challenge. This may have supplied another reason for Shakespeare to create Roderigo—Iago is base, thus not worthy to fight Cassio in open combat, as the Ensign fights the Captain in Cinthio's story:

> *Having formed this resolution, the Ensign, going out one dark night, sword in hand, met the Captain on his way to visit a courtesan, and struck him a blow on his right thigh, which cut off his leg and felled him to the earth. Then the Ensign was on the point of putting an end to his life, when the Captain, who was a courageous man and used to the sight of blood and death, drew his sword, and wounded as he was, kept on his defense, exclaiming with loud voice, "I'm murdered!" Thereupon the Ensign, hearing the people come running up, with some of the soldiers who were lodged thereabouts, took to his heels to escape being caught; then turning about again, he joined the crowd, pretending to have been attracted by the noise.* (385)

Though the Ensign surely shows he too is less than honorable with his ambush, Shakespeare intensifies Iago's cowardice by having Roderigo do the dirty work for Iago, thus avoiding even the possibility of defeat. By doing this, Shakespeare highlights Iago's base nature. Mirroring Cinthio's version, when the Ensign wounds the Captain in the leg, Shakespeare's stage direction is "*Iago from behind wounds Cassio in the leg, and exit*" (IV.ii.26), while Cassio is occupied with the dying Roderigo, something that makes Iago's cowardice far more prominent. Even the ethically villainous Suffolk stands forth against Whitmore's suffering death because "True nobility is exempt from fear" (*Henry VI, Part II*, IV.i.129), the difference here being that Suffolk is a degenerate gentle rather than a man of base blood, like Iago. *Peter Idle's Instructions to his Son* notes that it is "Manly to fight and not to flee as a cowarde."[45] As argued by Berkeley, "Shakespeare particularly faults his numerous villeins for lacking the classic virtue of courage (they are cowards),"[46] and *Othello*'s villain offers no contradiction to this statement.

Besides lacking courage, Iago is totally selfish, espousing a philosophy of egotism presented mostly in his soliloquies and in his dialogues with Roderigo. The source, since it presents little of the Ensign's speech, does not supply us with a direct comparison, so why does Shakespeare give Iago this quality of character? Theodore Spencer may have hit closest to the mark when he identifies Iago as "an unscrupulous individualist."[47] He feels Iago is compounded of three concepts of man's nature:

> *The difference between outer show and inner fact, the concept of the evil man as an individualist, and, connected with this, the concept of the evil man as the incomplete man, the man who does not contain all the psychological levels that should make up a human being.*[48]

The implication is that Spencer views Iago as a Machiavellian villain in direct conflict with the Renaissance idea of universal order, and that Shakespeare is reacting to the forces upon his own age.[49] Iago tries to impose his will on the natural order of things when he speaks:

> *Virtue? A fig! 'Tis in ourselves that we are thus or thus. Our bodies are our gardens, to the which our wills are gardeners; so that if we will plant nettles or sow lettuce, set hyssop and weed up thyme, supply it with one*

> *gender of herbs or distract it with many, either to have it sterile with idleness or manur'd with industry—why, the power and corrigible authority of this lies in our wills.* (I.iii.322–29)

Of course, the Renaissance view of God as being the controlling force in all things sees this individualism as a destructive problem. Brathwait even cautions about talking with oneself: "Take heed … thou talke not *with thine* Enemie,"[50] the enemy here being Satan.

Given this conflict, it is not hard to understand why Iago never repents, as does Othello at the end of the play—their differences in blood most likely are the key. Cinthio's ending leaves much to be desired dramatically. The Ensign "returns to his own country, and follow[s] up his wonted villainy." He is eventually tortured, dying "a miserable death" (Cinthio 388–89). Shakespeare's version is far more satisfying in a dramatic sense, with no loose ends as Lodovico orders "the censure of this hellish villain" (V.ii.377). Othello compares himself to a "base Indian" (V.ii.356) who throws away the jewel of Desdemona. Recalling his killing "a malignant Turk,"(V.ii.365), he repents in the only way left available to him—stabbing himself. But just before this he asks Iago to explain himself, giving the villain a chance to repent. Iago responds, "Demand me nothing. What you know, you know. From this time forth I never will speak a word" (V.ii.311–12). Lodovico incredulously asks, "What, not to pray?" (V.ii.313), but Iago, true to his word, speaks no more, even in the face of damnation.

The evidence that Shakespeare presents to us by way of the changes he makes from Cinthio's novel, intensifying the class distinctions, leads to the basic thesis of this essay: Iago *is* base in the class as well as in the ethical sense. Still, there are a few apparent incongruities present. Shakespeare only once in the canon allows a base character besides Iago to speak in blank verse, but Caliban is a special case—his mother is a witch, so is he even human? And can Caliban in any way be compared comfortably to Iago? There is also the matter of Iago's keen intelligence: though limited in scope, his mind sucked into immediacy, he is nonetheless a cunning, inventive man, so unlike any of Shakespeare's other base villains—at least at first glance. An easy resolution to these problems would be to say that Iago is a degenerate gentle, but the way he is put down in a class sense throughout the play, coupled with his acceptance of slurs, suggests otherwise. It is necessary to go beyond the boundaries of the play into a speculative argument that Iago is a bastard whose base blood is infused with some degree of gentility. This origin would explain his absolute ire when he does not receive the rank of gentleman that would have gone along with his commissioned status. Quite possibly, if the speculation is true, he even feels this rank is his birthright. Certainly Shakespeare presents other bastards in a similar way, such as Edmond in *King Lear*—an intelligent man, but flawed because of the limitations caused by his weak blood. Draper, in "Bastardy in Shakespeare's Plays," writes that "bastard was generally a term of reproach, as Shakespeare's use of it shows" and that "Edmund in his famous soliloquy tries to nullify the evil connotation of the word *base* with which the Elizabethans etymologically connected it."[51] There is also the question of why Othello, a noble man, even has a plebeian like Iago as his friend. There is, perhaps, a simple answer. Could Shakespeare have intended the play as a parody of the cult of male friendship, a subject Shakespeare deals with in many of his comedies, such as *The Merchant of Venice*? Peter Idle cautions against choosing the wrong company:

Looke also for ony disporte or othir thyng
With good felawshippe you be acompanyed;
That shall to honoure specially the bryng
If thou with hem woll be allied;
And he that with evell companye is asspied,
Ofte is takei n liknesse of a theiff
And fynally broght vnto myscheiff.[52]

Brathwait also offers this advice in *The English Gentleman*:

See here the love of good men one to another;
for amongst evill men can be no true friendship.
For it is the ayme of acquaintance that makes
it good or evil; as to insinuate onesselfe into
acquaintance for their owne ends, to wit, to profit
by it, or worke on others weaknesse, this is
acquaintance for Machiavells Schollers, whose
principall ayme is to undermine; and under pretence
of amity, shroud their villany.[53]

That Brathwait made reference to Machiavelli is especially informing, given Spencer's assertions about Iago. Whether Shakespeare had all these threads in mind when he wove the fabric of

Othello together is hard to say, though it certainly presents some compelling questions that could be developed in further studies.

The portrait of evil presented by *Othello* is varied and complex, and this study is by no means intended to be definitive. Yet it does offer a perspective on Shakespeare often avoided by critics who view his class bias with disdain or who, perhaps worse yet, dismiss the matter completely. But examining Shakespeare in this way allows us a clearer view of his possible intentions, especially the reasoning behind the extensive changes he made from his various source materials. All angles should be presented when looking at the multifaceted gem of Shakespeare's genius. On surer grounds one can say this with some degree of certainty: when reading *Othello* we can echo Brabantio's shout to Iago, "Thou art a villain," its double meaning intact.

Notes

1 *OED* cites the dates for this usage of the term *villain* as between 1303 and 1855 (A and B, 1–4).

2 By no means does the author here imply that the many critics who avoid this element present in the play render their discussions invalid. On the contrary, many apply directly to the points presented in this essay and will be cited when appropriate. Still, there are some, such as Stanley Edgar Hyman in his book *Iago: Some Approaches to the Illusion of His Motivation,* that go so far as to argue that Iago is a latent homosexual—perhaps an example of a modern critic's overreaching in his reading of the play; *Iago* (New York: Atheneum, 1970).

3 John W. Draper, *The Humors and Shakespeare's Characters* (Durham: Duke UP, 1945) 4.

4 C.S. Lewis, *Studies in Words,* 2nd ed. (Cambridge: Cambridge UP, 1960) 122.

5 David Shelley Berkeley, *Blood Will Tell in Shakespeare's Plays* (Lubbock: Texas Tech. Press, 1984) 7.

6 Draper and Berkeley have similar points of view in their readings of Shakespeare's plays; it would be prudent for any student of Shakespeare to consult these works in relation to this matter of Shakespeare's use of this physiological-based psychology.

7 Berkeley, *Blood* 8.

8 Robert B. Heilman, *Magic in the Web* (Lexington: U of Kentucky P, 1956) 111.

[9] Marvin Rosenberg, *The Masks of* Othello (Berkeley and Los Angeles: U of California P, 1961) 170.

[10] Harley Granville-Barker, *Preface to* Othello (Princeton: Princeton UP, 1958) 90.

[11] John W. Draper, *The* Othello *of Shakespeare's Audience* (New York: Octagon Books, 1966) 19.

[12] Draper, *Audience* 16.

[13] The translation from Italian is by John Edward Taylor. All of the quotations from the source are taken from the reprint of his 1855 translation in the Furness Variorium of *Othello* (Philadelphia: J.B. Lippincott, 1886).

[14] Quotations from *Othello* and other of Shakespeare's plays follow *The Complete Works of Shakespeare,* ed. David Bevington, 3rd ed. (Glenview: U of Chicago, 1980).

[15] Hoby was the first to translate Castiglione's text into English from the Italian. Translation used for this essay is that of Leonard Eckstein Opdycke.

[16] Count Baldesar Castiglione, *The Booke of the Courtier,* trans. Leonard Eckstein Opdycke (Immortal Classics, 1929) 11.

[17] Gerald M. Pincess and Roger Lockyer, *Shakespeare's World* (New York: Continuum, 1968) 155.

[18] Giovanni Baptista Geraldo Cinthio, *The Hecatommithi,* trans. Wolstenholme Parr. Rpt. in *The Variorum Shakespeare: Othello,* ed. Horace Howard Furness (Philadelphia: J.B. Lippincott, 1886) 387.

[19] Ifor Evans, *The Language of Shakespeare's Plays* (London: Methuen, 1964) 151.

[20] Heilman, *Magic* 105.

[21] Draper, *Audience* 21.

[22] Berkeley, *Blood* 14.

[23] Berkeley, *Blood* 48.

[24] Robert Burton, *The Anatomy of Melancholy,* ed. Floyd Dell and Paul Jordan-Smith (New York: Tudor, 1948) 82.

[25] Berkeley, *Blood* 54.

[26] Berkeley, *Blood* 29.

[27] Berkeley, *Blood* 50.

[28] Henry Peacham, *The Compleat Gentleman* (1622; Amsterdam: Da Capo Press, 1968) 185.

[29] Peacham, *Compleat* 186.

[30] Pincess and Lockyer, *World* 40.

[31] Pincess and Lockyer, *World* 41.

32 Pincess and Lockyer, *World* 43.

33 Richard Brathwait, *The English Gentleman* (London, 1630) 154.

34 Marianne Novy, *Love's Argument* (Chapel Hill and London: U of North Carolina P, 1984) 4.

35 Novy, *Argument* 4.

36 Heilman, *Magic* 106.

37 William Shakespeare, *Othello,* ed. Norman Sanders (Cambridge: Cambridge UP, 1985) 32.

38 Theodore Spencer, *Shakespeare and the Nature of Man,* 2nd ed. (New York: Macmillan, 1971) 133.

39 Thomas Cogan, *The Haven of Health* (London, 1606) 240.

40 Giovanni Nenna, *A Treatise of Nobility,* trans. W. Jones (London, 1595) 69.

41 Many modern critics see Iago as motiveless, thus fostering claims, such as Hyman's, that Iago is Satan.

42 George Whetstone, *The Honorable Reputation of a Souldier* (London, 1585) C.

43 Whetstone, *Souldier* Ciiij.

44 Sanders, *Othello* 25.

45 Peter Idle, *Instruction to His Son,* ed. Charlotte D'Evelyn (Boston: D.C. Heath, 1935) 95.

46 Berkeley, *Blood* 83.

47 Spencer, *Nature* 132.

48 Spencer, *Nature* 135.

49 Spencer sees Iago as representative of the Renaissance ideal of the man whose reason controls his passions. *Shakespeare and the Nature of Man* discusses the basis for his contention about Iago, that being the conflict between the Renaissance ideal of a universal order and the ideas of Machiavelli. While a few modern critics cite Iago as a Machiavellian villain, to my knowledge no critic has discussed this notion with regard to the class distinctions Shakespeare makes in *Othello.*

50 Brathwait, *English* 236.

51 Draper's discussion includes Shakespearean analogies: Don John in *Much Ado about Nothing*; Edmund and probably Oswald in *King Lear*; and possibly Thersites in *Troilus and Cressida.* All are examples of bastards "illegitimate both by fact of nature and by law" 127; "Bastardy in Shakespeare's Plays," *Shakespeare-Jahrbuch* 58 (1938): 123–136.

52 Idle, *Instruction* 83.

53 Brathwait, *English* 237.

Works Cited

Brathwait, Richard. *The English Gentleman.* London, 1630.

Berkeley, David Shelley. *Blood Will Tell in Shakespeare's Plays.* Lubbock: Texas Tech. P, 1984.

Bucknell, John Charles, M.D. *The Medical Knowledge of Shakespeare.* 1860. New York: AMS Press, 1971.

Burton, Robert. *The Anatomy of Melancholy.* 1621. Ed. Floyd Dell and Paul Jordan-Smith. New York: Tudor, 1948.

Castiglione, Count Baldesar. *The Booke of the Courtier.* Trans. Leonard Eckstein Opdycke. Immortal Classics, 1929.

Cinthio, Giovanni Baptista Geraldo. *The Hecatommithi.* Trans.Wolstenholme Parr. Rpt. in *The Variorum Shakespeare: Othello.* Ed. Horace Howard Furness. Philadelphia: J.B.Lippincott, 1886.

Cogan, Thomas. *The Haven of Health.* London, 1606.

Draper, John W. "Bastardy in Shakespeare's Plays." *Shakespeare-Jahrbuch* 58 (1938): 123–136.

___. *The Humors and Shakespeare's Characters.* Durham, North Carolina: Duke UP, 1945.

___. *The* Othello *of Shakespeare's Audience.* New York: Octagon Books, 1966.

Elliott, G.R. *Flaming Minister.* Durham, North Carolina: Duke UP, 1953.

Evans, Ifor. *The Language of Shakespeare's Plays.* London: Methuen, 1964.

Granville-Barker, Harley. *Preface to* Othello. Princeton: Princeton UP, 1958.

Heilman, Robert B. *Magic in the Web.* Lexington: U of Kentucky P, 1956.

Hyman, Stanley Edgar. *Iago.* New York: Atheneum, 1970.

Idle, Peter. *Instruction to His Son.* Ed. Charlotte D'Evelyn. Boston: D.C. Heath, 1935.

Lewis, C.S. *Studies in Words.* 2nd ed. Cambridge: Cambridge UP, 1960.

Milles, Thomas. *The Catalogue of Honor.* London, 1610.

Nenna, Giovanni. *A Treatise of Nobility.* Trans. W. Jones. London, 1595.

Novy, Marianne. *Love's Argument.* Chapel Hill: U of North Carolina P, 1984.

Peacham, Henry. *The Compleat Gentleman.* 1622. Amsterdam: Da Capo Press, 1968.

Pincess, Gerald M., Roger Lockyer. *Shakespeare's World.* New York: Continuum, 1989.

Rosenberg, Marvin. *The Masks of* Othello. Berkeley: U of California P, 1961.

Shakespeare, William. *The Complete Works of Shakespeare.* Ed. David Bevington. 3rd ed. Glenview: U of Chicago, 1980.

___. *Othello.* Ed. Norman Sanders. Cambridge: Cambridge UP, 1985.

___. *Othello.* Ed. Horace Howard Furness. Philadelphia: J.B. Lippincott, 1886.

Spencer, Theodore. *Shakespeare and the Nature of Man.* 2nd ed. New York: Macmillan, 1971.

Whetstone, George. *The Honorable Reputation of a Souldier.* London, 1585.

The Art of Class Delineation: The Aesthetic Disparity Between *The Shrew* and *A Shrew*

by Kurt Hochenauer

The courtship of Kate and Petruchio in *The Taming of the Shrew* mirrors the prevailing hierarchy between men and women in Shakespeare's society, suggests the importance of patriarchal influence in Elizabethan marriages, and reconciles romantic love with the practical business of an arranged marriage among the gentility. The play reminds us, perhaps above all else, that Shakespeare reinforces conservative Elizabethan social-class codes. Thus, Kate admonishes Petruchio, "So may you lose your arms / If you strike me, you are no gentleman / And if no gentleman, why then no arms."[1] This is a telling moment in *The Shrew*. Kate's punning argument places her relationship to Petruchio under moral guidelines held by the armigerous in Elizabethan society. Her response also reflects the class-conscious framework of the taming plot in *The Shrew*, a framework missing from *The Taming of a Shrew* (hereafter referred to as *A Shrew*), published anonymously in 1594. In *A Shrew*, for example, Kate never once mentions the coat of arms held by Petruchio's counterpart, Fernando. It is this difference and numerous others, based on Shakespeare's preoccupation with maintaining a strict division between classes, that account ultimately for the aesthetic disparity between the two plays.

In this regard, Shakespeare's plays do not reflect an isolated, Elizabethan philosophy. His plays suggest, for example, that Shakespeare adopts the Count's bifurcation of humans in *The Book of the Courtier*, first published in English in 1561:

> *It is true that, by favor of the stars or of nature, some men are endowed at birth with such graces that they seem not to have been born, but rather as if some god has formed them with his very hands and adorned them with every excellence of mind and body. So too there are many men so foolish and rude that one cannot think that nature brought them into the world of contempt or mockery.* [2]

Those characters with excellence of mind and body make up the vast majority of the gentility in Shakespeare's plays. Those characters foolish and rude are most often base-born. In addition, Shakespeare's plays cling to "the dominant Elizabethan view that heredity ... determines human individuals."[3] Thus, Ulysses in *Troilus and Cressida* may well be speaking for Shakespeare when he lauds the importance of degree in "primogenity and due of birth" (I.iii.106). *The Shrew*, published in the 1623 First Folio and one of the first comedies, follows these precepts, and Shakespeare uses them to create a better play than *A Shrew*. Many critics, perhaps because of their own modern democratic ideologies, ignore the fact that the aesthetic difference between *The Shrew* and *A Shrew* not only shows Shakespeare's brilliance but also reveals an underlying Elizabethan philosophy that stresses the rigid separateness of two classes, armigerous and base. *The Shrew's* Induction scenes, for example, are longer, and no reader would deny that Sly, a base-born character, becomes a developed and hilarious character in Shakespeare's hands. Yet the heightened humor comes at the expense of Sly's foolishness and rudeness. When the Lord tells Sly he must call his wife "madam," Sly asks, "Al'ce madam or Joan madam?" (Induction ii.107). Thus the humor derives from Sly's churlish stupidity and from his overall characterization as a base-born without social graces. This comic exchange between Sly and the Lord is not found in *A Shrew*.

Critics often apply psychological readings to *The Shrew*, ignoring that Kate and Petruchio's fictional courtship took place in a pre-Freudian era much different from our own. Thus, Harold Bloom,[4] Juliet Dusinberre,[5] and Coppelia Kahn[6] find that Kate's eventual acquiescence is ironic. Other critics of *The Shrew* note that Shakespeare stresses an ambiguous view of the patriarchal order,[7] or else presents his unsympathetic view of a bourgeois marriage.[8] What most of these critics have in common is a desire to give Shakespeare's play an egalitarian reading based on what they perceive as the play's ironies. Moreover, no certain textual or historical evidence exists or has been found that solves the relationship between *The Shrew* and *A Shrew* to everyone's satisfaction. Obviously the plays are related, as seen in the correspondences between plot and language, but for more than a century, Shakespearean scholars have remained divided over how to account for the later publication date of *The Shrew*. Further, the staging history of the two plays is uncertain.

This study will not attempt to solve the publishing relationship between the plays, a task filled always with uncertainty and speculation.[9] Instead, this study provides a comparison of the plays which reveals that Shakespeare's brilliance in *The Shrew* is entrenched in a conservative, Elizabethan philosophy that compels the author to heighten class differences, reinforce existing class structure among his characters, and focus on characters' heredity.[10] Comparing the two plays, I shall show how the increased awareness of class consciousness and heredity present in *The Shrew* permeates the differences in the two plays' Induction scenes, in Kate and Petruchio and their relation to the taming plot, in the master-servant relationship between Petruchio and Grumio, and in the circumstances of three other primary characters, Lucentio, Bianca, and Baptista. These differences enhance the comedic elements in *The Shrew*, and, in the case of the taming plot, the differences serve to unify the play. The primary characters are more developed, well-rounded, subtle, and thematically consistent through Shakespeare's amplifications of their gentility. Overall, the differences show, as John W. Draper comments about Shakespeare generally, "a greater truth of character, appropriately expressing itself in the related actions of the plot."[11]

The Induction scenes in both *A Shrew* and *The Shrew* serve as an excellent starting point because they present a microcosm of the division between gentle and base. The plot of the inductions in both plays is similar: a lord finds a sleeping drunk on his property, gives him fine food and drink, and then convinces him to watch the play. In *A Shrew*, when the Lord finds Slie on his property, he says: "Fie, how the slavish villaine stinkes of drinke" (i.24). In *The Shrew*, the Lord's reaction is longer and more poetic, suggesting the usual

Shakespearean relationship between gentle and base: "O monstrous beast, how like swine he lies! / Grim death, how foul and loathsome is thine image" (Induction i.33–34). Shakespeare not only adds a line but heightens the rhetoric's quality as well. Both lords then describe how they plan to dupe the drunken rogue. In *A Shrew*, the Lord tells his servants:

Go take him uppe beare him to my house,
And beare him easilie for feare to wake,
And in my fairest chamber make a fire
And set a sumptuous banquet on the boord
And put my richest garments on his backe,
Then set him at the Table in a chaire:
When that is doone against he shall awake,
Let heavenlie musicke play about him still. (i.26–33)

In *The Shrew*, Shakespeare adds sensuous detail that corresponds to the Lord's social position:

Carry him gently to my fairest chamber
And hang it round with all my wanton pictures,
Balm his foul head in warm distilled waters
And burn sweet wood to make the lodging sweet.
Procure me music ready when he wakes,
To make a dulcet and a heavenly sound. (i.45–51)

Such lush detail here and throughout the Induction scenes in *The Shrew* serves to widen the social gap between Sly and the Lord because it reinforces the differences in language and material objects between the base and armigerous. This becomes more apparent when Sly wakes up and asks: "For God's sake, a pot of small ale" (Induction ii.1). There is a parallel, almost identical, line in *The Shrew.* Small ale is obviously used as a class discriminant in both plays. But Shakespeare makes much more of Sly's base qualities in the ensuing passages, and he builds the comic effect using food and drink. In *The Shrew,* Sly is offered wine and conserves. But he admits: "I ne'er drank sack in my life; and if you give me any conserves, give me conserves of beef" (Induction ii.6–7). There is no related passage in *A Shrew.* Wine is the preferred beverage of the gentility, serving to replenish and warm a person's blood supply. Falstaff notes this in *2 Henry IV* when he attributes Prince Hal's warm blood to the drinking of sack (IV.iii.116–21). Beef is often associated with the base in Shakespeare's plays. Thus, Thersites uses the epithet "beef-witted" in *Troilus and Cressida*

(II.i.12–13). Therefore, the fact that Sly refused the wine and asks for beef shows his base qualities. The humor is heightened in *The Shrew*, and it comes at the expense of a plebeian placed out of his base element, a common Shakespearean tactic in the comedies. In *A Midsummer Night's Dream*, for example, Quince and his crew, all workers or craftsmen, perform a short play that, among other things, shows their ineptness once placed out of their day-to-day worlds. As they attempt to appropriate identities associated with art and rhetoric—the territory of the armigerous—the result is humor at their expense.

In *The Shrew*, Sly actually fights for his right to be base, protesting to the Lord that "Am I not Christopher Sly, old Sly's son of Burtonheath, by birth a pedlar, by education a card maker, by transmutation a bear herd, and now by profession a tinker?" (Induction ii.17–20). The Lord responds: Call home thy ancient thoughts from banishment / And banish hence these abject lowly dreams" (Induction ii.31–32). Thus, Sly's class and heredity remain the focus of the comedy in the scene. Again, there are no related passages in *A Shrew*. This focus provides an example of how Shakespeare, aligned with Elizabethan views of heredity, uses a character's lineage to expand and improve *The Shrew*. A few lines later, Sly becomes convinced of his new social status, and asks the Lord to bring his supposed wife into the room. Yet even here Sly shows his baseness when he adds "And once again, a pot o' the smallest ale" (Induction ii.75). In *A Shrew*, Slie says about his wife, "For she and I will go to bed anon" (ii.45). In *The Shrew*, Sly, of course, is more direct and vulgar: "Tis much. Servants leave me and her alone. Madam, undress you and come now to bed" (Induction ii.113–14). Sly's vulgarity here and elsewhere in the Induction, juxtaposed against the Lord's gentility, makes Sly a more comic character.

The plays' taming plots provide more evidence that Shakespeare relies on conservative Elizabethan beliefs about matters of class and lineage to create a better play than *A Shrew*. In *The Shrew*, for example, the taming plot is conducted under a class-consciousness lacking in *A Shrew*. When we are first told about Kate in Shakespeare's play, we learn she is "With wealth enough, and young and beauteous, / Brought up as best becomes a gentlewoman" (I.ii.85–86). In *A Shrew*, Kate is sometimes described as "lovelie" by a jesting Fernando, Petruchio's counterpart, but she is never given the gentle qualities

Shakespeare gives her. Again, lineage, more than wealth, remains important in *The Shrew.* In this regard, Shakespeare reflects Henry Peacham's view in the 1622 *The Compleat Gentleman*: "Neither must we honor or esteem those ennobled, or made gentle in blood, who by ... base means have raked up a mass of wealth."[12] *The Shrew*'s Petruchio, for example, describes his lineage by mentioning his father first:

Antonio, my father, is decease'd,
And I have thrust myself into this maze,
Happily to wive and thrive as best I may.
Crowns in my purse I have and goods at home,
And so am come abroad to see the world. (I.ii.53–57)

In *A Shrew,* we know Fernando has "wealth sufficient" (iv.48), but his lineage remains a mystery. Shakespeare, on the other hand, ensures that the audience is made aware that both Petruchio and Kate have the proper bloodlines. This is important because, as in the Induction scenes, Shakespeare heightens the humor of *The Shrew*'s taming plot through social-class distinctions. I have already mentioned, for example, Kate's reference to Petruchio's coat of arms, a punning reference designed to amuse an Elizabethan audience. But it also reflects how Shakespeare intertwines a class consciousness with the play's comic elements.

In *The Shrew,* as well, Shakespeare lengthens the exchange between Kate and Grumio about food. He does so, as Berkeley suggests, to perhaps reinforce that Kate has fallen so far from her gentle qualities she can now be linked to the vulgar Sly in the Induction.[13] When Grumio offers her beef and mustard, Kate tells him that it is "A dish that I do love to feed upon" (IV.iii.25). Earlier, Petruchio refuses to give Kate meat because "'twas burnt and dried away, / And I expressly am forbid to touch it, / For it engenders choler, planteth anger'" (IV.i.158–60). This reflects the Elizabethan idea that dried meat, in particular, upsets a person's humoral balance. "Generally, [meats] should bee liquid and informe of brothes," writes Timothy Bright in the 1586 *A Treatise of Melancholie.*[14] Shakespeare heightens the comic effect by using standard Elizabethan views about diet.

Later, Kate uses class terms to express her appreciation of the haberdasher's cap: "I'll have no bigger. This doth fit the time, / And gentlewomen wear caps such as these" (IV.iii.69–70). Petruchio, however, refuses her the cap and places their stormy relationship in a class-conscious framework.[15] "When you are gentle, you shall have one too, / And not till then" (IV.iii.71–72). The word "gentle" holds a double meaning, punning on Kate's previous reference to gentlewomen, and it is designed undoubtedly to amuse an Elizabethan audience. By contrast, Fernando in *A Shrew* tells Kate he will allow her to return to her father's house only "when you're meeke and gentell but / Before, I know your stomack is not yet come down" (xi.43–44). The class connotations, along with the humor, are obviously lacking in *A Shrew.*

The difference in Kate's final speech in the two plays reinforces Shakespeare's intensification of the class consciousness framework of *The Shrew.* In *A Shrew,* Kate's admonishment to the women is based solely on prevailing religious codes. In *The Shrew,* however, Kate unifies the class consciousness theme by placing the man/woman relationship in class terms: "Thy husband is thy lord, thy life, thy keeper, / Thy head, thy sovereign; one that cares for thee" (V.ii.148–49). Kate continues the analogy a few lines later: "Such duty as the subject owes the prince / Even such woman oweth to her husband" (V.ii.157–58). If a woman is not obedient, she is a "graceless traitor to her loving lord" (V.iii.162). Conspicuously absent are the glaring religious terms set forth by *A Shrew*'s Kate. As Ann Thompson points out, "Shakespeare does not mention God at all in his version of this speech. His view of marriage is purely a secular one."[16] Those religious terms in *A Shrew* reflect common beliefs embraced by the play's audience. They also show a lack of creativity. Using this well-worn argument, *A Shrew*'s author relies on conventional material. Shakespeare, on the other hand, not only uses different analogies but also brings his taming plot full circle by using his class awareness. Kate's comments about Petruchio's coat of arms foreshadows her final speech. The taming plot, then, is associated consistently with class matters, from the time Kate and Petruchio first meet to Kate's final display of submission. This helps give the play a unity lacking in *A Shrew.* Moreover, Kate's submission maintains the prevailing patriarchal order of the play.

Given Shakespeare's focus on class and lineage, it makes perfect sense that he also heightens the humor of *The Shrew* through the relationship between master and servant. This is reflected the first time we encounter Petruchio and Grumio in *The Shrew.* Petruchio lambasts Grumio for refusing to knock on Hortensio's door. "Villain, I say, knock me at this gate, / And rap me well, or I'll knock your knave's pate" (I.ii.11–12). There is no such scene in *A Shrew,* and again I point out that the humor is at the expense of a base character. This reflects a standard subplotting device Shakespeare employs in several plays, including *The Comedy of Errors.* Dromio of Syracuse, for example, already beaten severely by his master, asks, "Master, shall I be porter at the gate?" (II.ii.216), to which Adriana responds, "Ay, and let none enter, lest I break your pate" (II.ii.217). This similarity between passages in *The Shrew* and *The Comedy* shows how Shakespeare commonly applies stock expressions to amplify the humor and the division between gentle and base. In addition, Shakespeare, maintaining the status quo in class matters, eliminates Saunder's offer of his sister to Fernando in *A Shrew*: "And I have a prettie wench / To my sister, and I had thought to have perfered my / Maister to her ..." (v.89–91). Such a match, even if proposed in jest, would perhaps be repugnant to Shakespeare, and thus the baseborn Grumio never makes such an outlandish and implausible offer to his master, Petruchio.

There remain the differences among the play's other principal characters. Typically, Lucentio gives his lineage in Act I of the play:

> *Pisa, renowned for grave citizens,*
> *Gave me my being and my father first,*
> *A merchant of great traffic throughout the world,*
> *Vincentio, come of Bentivolii.*
> *Vincentio's son, brought up in Florence,*
> *It shall become to serve all hopes conceiv'd,*
> *To deck his fortune with his virtuous deeds.* (I.i.10–16)

By contrast, his counterpart in *A Shrew,* Aurelius, a duke's son, disguises himself as a merchant's son. "But when we come unto hir fathers house, / Tell him I am a Marchants sonne of Cestus / That comes for trafficke unto Athens heere" (iv.57–59). By making Lucentio a merchant's son and Bianca a merchant's daughter, Shakespeare stresses the prevailing status quo in class matters.[17] More importantly, he adds a great deal of detail about Lucentio's background that stresses Lucentio's gentility. We know more about Lucentio's gentility in *The Shrew* than we do about Aurelius's royal background in *A Shrew.*

The detail underlining the characters' social class is lacking in the earlier play.

But given Shakespeare's concern with lineage, it is not surprising that Lucentio's father is a great merchant whereas Aurelius's father is a duke. When the Duke in *A Shrew* finds out his son has married a merchant's daughter, he expresses his outrage: "Alfonso I did not thinke you would presume / To match your daughter with my princely house / And nere make me acquainted with the cause" (xvi.64–66). Shakespeare eliminates this tension. In Shakespeare's plays, characters have degrees of blood; a royal family would obviously have "higher" blood than a merchant family. To match a duke's son with a merchant's daughter, while not completely improbable within Shakespeare's plays, does stretch the laws of probability. In *All's Well That Ends Well*, for example, the noble Bertram goes to great extremes to annul his marriage to Helena, a physician's daughter and a gentlewoman. *A Shrew*'s Aurelius, of course, exhibits none of Bertram's concern about mingling his blood with that of a woman lower in rank than he. In addition, Aurelius's father, though he protests his son's marriage in class terms, is quickly and improbably appeased: "Aurelius stand up I pardon thee" (xvi.103). He then extends his pardon to Bianca's counterpart, Phylena. "And you faire virgin too I am content, / To accept you for my daughter since 'tis done" (xvi.107–8). We might assume, given Shakespeare's ideology and artistic talent, that the playwright would have found the Duke's easy acceptance as implausible as the marriage itself. Consequently, he recast the marriage between Bianca and Lucentio.

Of *The Shrew*'s Bianca, we learn she possesses the qualities of good blood because "she taketh delight / In music, instruments, and poetry" (I.i.92–93). In addition, Lucentio describes her beauty in terms commonly associated with the gentility in Shakespeare's plays: "Tranio, I saw her coral lips to move, / And with her breath she did perfume the air. / Sacred and sweet was all I saw in her" (I.i.174–76). *A Shrew*'s Phylena, Bianca's counterpart, is described in glowing terms as well by Aurelius. She is "the image of honor and nobilitie" (iv.33), and she has a "sacred beautie" (iv.53). But the description lacks the depth and detail of Lucentio's smitten exuberance: "O yes, I saw sweet beauty in her face / Such as the daughter of Agenor had, / That made great Jove to humble him to her hand / When with his knees he kiss'd the Cretan strand" (I.i.167–70). Unlike Phylena, Bianca, then, holds all the good qualities of a gentlewoman, including sweet-smelling breath, a trait Shakespeare repeatedly attributes to the gentility in his plays. In *The Merchant of Venice,* Bassanio, for example, describes Portia's lips as "parted with sugar'd breath" (III.ii.119). By contrast, baseborn people in Shakespeare's plays often have bad breath, as evidenced by Casca's comment about a crowd of plebeians who "uttered such a deal of stinking breath" in *Julius Caesar* (I.ii.247). Bianca possesses this among other traits that mirror gentility throughout the canon. These physical traits that show social class are what Shakespeare often intensifies in his characters. Bianca also shows the proper respect, like most gentility, for her family, including her sister. After Kate binds her hands, for example, Bianca tells her: "Unbind my hands, I'll pull them off myself, / Yea, all my raiment, to my petticoat, / Or what you will command me will I do, / So well I know my duty to my elders" (II.i.4–7). There is no corresponding passage in *A Shrew,* and few base characters in Shakespeare's canon are given this strong sense of family loyalty. Bianca's strong sense of family duty here makes her final (and slight) disobedience to her husband all the more poignant.

Baptista is another character who benefits from Shakespeare's artistic brilliance and his concern with social-class codes. In *A Shrew,* Fernando is promised six thousand crowns by Alfonso, Baptista's counterpart, for marrying Kate. We are told of the arrangement by Fernando. In *The Shrew,* Baptista promises Petruchio "After my death the one half of my lands / And in possession twenty thousand crowns" (II.i.121–22). Shakespeare gives Baptista the lines, and he raises the rhetoric as well as the ante. Moreover, unlike Alfonso, Baptisa is concerned with the parentage of his daughter's suitors, reflecting, again, the Elizabethan idea that heredity more than wealth determines a person's social status. When Tranio, disguised as Lucentio, tells Baptista, "That, upon knowledge of my parentage, / I may have welcome amongst the rest that woo" (II.i.95–6), Baptista bases his welcome on Lucentio's lineage, "A mighty man of Pisa. By report / I know him well. You are welcome sir" (II.i.104–5). As a gentle created according to Shakespeare's code, Baptista is concerned foremost with the suitor's parentage. Once satisfied that the good blood of his daughters will not mingle with base blood, Baptista gets down to the business of dowries. In *A Shrew,* Alfonso does not concern himself with the suitor's family background. In *The Shrew*, after Baptista finds out his daughter has married without his consent, he calls it "the depth of knavery" (V.i.129). Alfonso, after he discovers his daughter has married a Duke's son, is not overly concerned that he has personally been duped. Instead, he simply begs the Duke's forgiveness and promises that "I will in humblest manner I can, show / The eternall service I do owe your grace" (xvi.14–15). This difference in the plays, again, represents the typical manner in which Shakespeare amplifies the social-class codes in his plays. It shows how Shakespeare's character development—Baptista has a significant role in Shakespeare's play, and his patriarchal status is never really questioned—is connected often to Elizabethan social-class issues.

Almost without exception, the aesthetic improvements made by Shakespeare in the Induction scenes, the taming plot, and the principal characters are connected in some way to class matters or matters of lineage. If a principal character holds a high station in life and has not degenerated, Shakespeare almost always amplifies that character's gentle traits. If a principal character is base, Shakespeare stresses his or her vulgarity. This is an artistic device, and it holds true with the relationship between *The Shrew* and *A Shrew*. Critics, in their quest to solve the relationship between the plays, often point out the aesthetic differences between them in general terms. They have long recognized that in Shakespeare's play the characters are more developed, the language is superior, and the comedy is more intense. But they have overlooked the conservative Elizabethan social codes underlying the aesthetic differences. It would be foolish to suggest that Shakespeare uses these codes as mere propaganda in *The Shrew*, but the themes do underscore the brilliant improvements he made in the play.

Further, if *The Shrew* is a representative example, and I think it is, then Shakespeare's writing was entrenched in an Elizabethan philosophy often dismissed by some critics who perhaps hope to see their own beliefs reflected in the plays. These critics may well be giving Shakespeare the greatest form of flattery by seeing their own ideologies reflected in the plays. Unfortunately, their exuberance is sometimes based on a distortion of Elizabethan reality.

Notes

1 *The Shrew* in *The Complete Works of Shakespeare,* Ed. David Bevington (Glenview, IL: Scott, Foresman, 1980), II.i.221–23. Further citations from Shakespeare are from this edition. For *A Shrew,* I use the version in *Narrative and Dramatic Sources of Shakespeare,* Vol. 1, ed. Geoffrey Bullough (New York: Columbia UP, 1957) 69–108. Lineation and quotations follow this text. I retain *A Shrew*'s original spelling.

2 Count Baldesar Castiglione, *The Book of the Courtier,* trans. Leonard Eckstein Opdycke (NP: Immortal Classics, 1929) 22.

3 David Shelley Berkeley, *Blood Will Tell in Shakespeare's Plays,* (Lubbock: Texas Tech P, 1984) 11.

4 Harold Bloom, Introduction, *William Shakespeare's* The Taming of the Shrew (New York: Chelsea, 1988) 4.

5 Juliet Dusineberre, *Shakespeare and the Nature of Women* (New York: MacMillan, 1975) 105.

6 Coppelia Kahn, "Coming of Age: Marriage and Manhood in *The Taming of the Shrew,*" William Shakespeare's *The Taming of the Shrew,* ed. Harold Bloom (New York: Chelsea, 1988) 41.

7 Marianne L. Novy, "Patriarchy and Play in *The Taming of the Shrew,*" William Shakespeare's *The Taming of the Shrew,* ed. Harold Bloom (New York: Chelsea, 1988) 13.

8 Carol F. Heffernan, "*The Taming of the Shrew*: The Bourgeoisie in Love," William Shakespeare's *The Taming of the Shrew,* ed. Harold Bloom (New York: Chelsea, 1988) 65.

9 Critics have posited three primary arguments: (1) *A Shrew* is a principal source for *The Shrew*; (2) *A Shrew* is a memorial reconstruction or bad quarto of *The Shrew*; and (3) *A Shrew* and *The Shrew* derive from a common and lost play, an *Ur-Shrew.* The latter two theories remain the most popular in modern criticism. For *A Shrew* as a principal source, see John W. Schroeder, "*The Taming of a Shrew* and *The Taming of the Shrew*: Case Reopened," *Journal of English and Germanic Philology* 57 (1958): 424–43; E. K. Chambers, *William Shakespeare: A Study of Facts and Problems,* vol. 1 (London: Oxford UP, 1930); Frederick Boas, *Shakespeare and his Predecessors* (New York: Scribner's, 1896); Ernest Kuhl, "The Authorship of *The Taming of the Shrew,*" *PMLA* 40 (1925): 551–618; and Thelma Greenfield, "The Transformation of Christopher Sly," *Philological Quarterly* 33 (1954): 34–42. For the bad quarto theory, see Samual Hickson, "*The Taming of the Shrew,*" *Notes and Queries* 1 (1850): 345–47; P. Alexander, "*The Taming of a Shrew,*" *Times Literary Supplement* (16 Sept. 1926): 614; B.A.P. Van Dam, "The Taming of a Shrew," *English Studies* 10 (1928): 96–106; John Wilson

and Sir Arthur Quiller-Couch, Afterword, *The Taming of the Shrew* (London: Cambridge UP, 1968): 110–26; J.C. Smart, *Shakespeare, Truth and Tradition* 1928 (London: Oxford UP, 1966); Henry David Gray, "*The Taming of a Shrew,*" Philological Quarterly 20 (1941): 325–33; E.A.J. Honigman, "Shakespeare's Lost Source Plays," *Modern Language Review* 49 (1954): 293–307; Richard Hosley, "Sources and Analogues of *The Taming of the Shrew,*" *Huntington Library Quarterly* 27 (1964): 289–308; and J.C. Maxwell, "*The Shrew* and *A Shrew*: The Suitors and the Sisters," *Notes and Queries* 213 (1968): 130–31. For the Ur-Shrew theory, see Albert Tolman, "Shakespeare's Part in *The Taming of the Shrew,*" *PMLA* 5 (1890): 201–77; Hardin Craig, "*The Shrew* and *A Shrew,*" *Elizabethan Studies and Other Essays,* vol. 2 (Boulder: U of Colorado P): 150–54; Raymond Houk, "The Evolution of *The Taming of the Shrew, PMLA* 57 (1942): 1009–38; W.W. Greg, *The Shakespeare First Folio: Its Bibliographic and Textual History* (New York: Oxford UP, 1955); and G.I. Duthie, "*The Taming of a Shrew* and *The Taming of the Shrew,*" *The Review of English Studies* 19 (1943): 337–56.

10 It should be noted that Shakespeare, stressing division between his two classes, often departs from his primary sources in this way. This pattern is repeated consistently throughout the canon. In *2 Henry IV,* for example, Shakespeare presents Jack Cade as a base ruffian who would "make it a felony to drink small beer" (IV.ii.65–66). Cade is presented as an unschooled, violent, power-hungry churl, and no credence is given to the group's legitimate concerns with the country's entrenched political corruption. By contrast, Raphaell Holinshed, *Holinshed's Chronicles of England, Scotland and Ireland,* 1587, vol. 3, 2nd ed., (New York: AMS Press, 1965), whom Shakespeare used as a source, presents a more balanced account. Holinshed notes that the nobles who tried to negotiate with Cade found him "sober in talke, wise in reasoning, arrogant in hart, and stiffe in opinion" (225). In *The Winter's Tale,* Shakespeare departs from his primary source, Robert Greene's *Pandosto: The Triumph of Time,* rpt. in *A New Variorum Edition of Shakespeare: The Winter's Tale,* ed. Horace Howard Furness (Philadelphia: Lippincott, 1898), 325–52. Shakespeare invents the episode in which Antigonus is attacked by a bear as the clownish shepherd's son stands idly by; this emphasizes the son's cowardly, base qualities and heightens the contrast between Perdita and the shepherd family. He also eliminates Pandosto's (Leontes') incestuous, base sexual desire for his daughter. (Shakespeare's abhorrence of incest may well be reflected by Gower in *Pericles,* Prologue.i.27–28: "Bad child, worse father, to entice his own / To evil should be done by none!") In *Henry V,* Shakespeare gives no credit to the archers who helped win the battle at Agincourt. Shakespeare ignores Holinshed's account that describes the English archery attack on the French footmen as an important turning point in the battle (81). We can assume Shakespeare

believed archery to be an ungentlemanly combat tactic, and thus he refused to include it in his play. Gentlemen, under Shakespeare's code, would engage in hand-to-hand combat. These deletions, additions, changes, and amplifications are representative of Shakespeare's treatment of his primary sources.

11 John W. Draper, "Country and Court in Shakespeare's Plays," *Stratford to Dogberry* (U of Pittsburgh P, 1961) 2.

12 Henry Peachman, *The Compleat Gentleman,* (1622; New York: DeCapo, 1968) 3.

13 Berkeley, *Blood,* 66.

14 Timothy Bright, *A Treatise of Melancholie* (London: NA, 1586) 313.

15 For an interesting account of how Shakespeare uses dress in *The Shrew,* see Kevin L. Seligman's "Shakespeare's Use of Elizabethan Dress as a Comedic Device in *The Taming of the Shrew*: Something Mechanical Encrusted on the Living," *Quarterly Journal of Speech* 60 (1974): 39–44. I disagree, however, with Seligman's conclusion that Shakespeare uses fashion to satirize "the foibles of an age rich in extravagant disguise ... " (39). After Kate and Petruchio become reconciled, it can be assumed that both wear the typical, flowery dress of the gentility.

16 Ann Thompson, Introduction, *The Taming of the Shrew* (Cambridge UP, 1984) 29.

17 Shakespeare uses George Gascoigne's *Supposes,* first published in English in 1575 and based on *I Suppositi,* as a source for the Lucentio-Bianca story. Shakespeare may be referring to the play when Lucentio tells Baptista: "That have by marriage made thy daughter mine, / While counterfeit supposes blear'd thine eyne" (V.i.109–10). Here again, however, my argument holds true. Shakespeare increases the division between classes in the Lucentio/Bianca story and creates a better play. Shakespeare, for example, eliminates the loss-of-virginity plot in *Supposes.* Bianca remains the gentlewomanly virgin throughout *The Shrew.* In addition, Tranio's counterpart in *Supposes* is discovered to be a gentleman. In *The Shrew,* Tranio's father remains "a sailmaker in Bergamo" (V.i.72). Shakespeare refuses to gentle Tranio, just as he refuses consistently to gentle other base people in his plays, with the notable exception in *Henry V.* In this play, King Henry gentles the English soldiers of Agincourt. "For he today that sheds his blood with me / Shall be my brother; be he ne'er so vile, / This day shall gentle his condition" (IV.iii.61–63). Shakespeare obviously felt obligated to hold true to his primary source material in this case.

Shakespeare's Villeinizing of Jack Cade

by Byung-Eun Lee

The scenes of Jack Cade's rebellion are the most vigorous and ridiculous scenes in *Henry VI, Part II.* Jack Cade and his followers rebel against the gentles, killing many aristocrats and destroying parts of London. Even though the rebels in the play insist that they desire to make all men equal and to have all things in common and thus might seem to desire a return to early Christian communism, they are as a class, in fact, ignorant and vain social climbers.

Some critics, Charles Hobdays[1] and Michael Hattaway,[2] for example, see the Jack Cade scenes as evidence of Shakespeare's grasp of the tradition of egalitarianism in post-fourteenth century England, which was based, like the Peasants' Revolt of 1381, on the idea that all Adam's descendents are equal. J. P. Brockbank[3] and M. M. Reese[4] also claim that even in the Cade episodes Shakespeare never loses his rich human sympathy. But Shakespeare is not at all sympathetic with Cade. Other critics—Brents Stirling,[5] Richard Wilson,[6] and Karl Brunner,[7] for example—see that Shakespeare, a landowner and (many think) a would-be gentleman,[8] naturally sided with the conservative element.

Stirling, pointing out that Shakespeare's account of the rebellion actually runs counter to that in Raphael Holinshed's chronicles, says that Shakespeare makes Cade "a ridiculous equalitarian and Utopian."[9] Richard Wilson eloquently insists that Cade is "metamorphosed into a cruel, barbaric lout, whose slogan is 'kill and knock down,' and whose story as 'the archetype of disorder' is one long orgy of clownish arson and homicide fuelled by an infantile hatred of literacy and law."[10] Brunner mentions that Shakespeare, in order to emphasize the "disruptive, vaguely communistic ideas and slogans of the crowd and the disasters which may come from a weak and divided government,"[11] tones down the fact that in the chronicles the King's opponents urge Cade on. Irving Ribner and George Lyman Kittredge[12] also state that Shakespeare altered the character of Cade in Holinshed to make Cade a vice character in the morality play tradition, yet even Stirling, Wilson, Brunner, Ribner, and Kittredge quite overlook distinctions between gentle and base blood with which Shakespeare in the Cade episodes is very much concerned.

Blood was anciently regarded as determining its owner's appearance, mentality, and character. Thomas Walkington mentions that a person's soul is contained in his blood,[13] a belief also seen in Andrew Boorde's reference to "good blode, in the whiche consysteth the lyfe of man."[14] Social classification according to the quality of blood, as distinguished from an appraisal of human beings on egalitarian principles, was the predominant Elizabethan view, and certainly it was the view of Tudor political conservatives. Shakespeare, as David S. Berkeley says, was a rigid defender of this belief and divided his dramatic world into two blood-based classes, the gentle and the base. In Shakespeare's plays the gentle, if not degenerate, are almost always rewarded by being given positions that accord with their blood quality, while the base are usually humiliated or destroyed if they aspire to climb to a higher social station. Shakespeare's principle operates in *2 Henry VI.* As occasionally observed by critics, Shakespeare in *2 Henry VI* melded, from the chronicles of Holinshed and Edward Hall, the Peasants' Revolt of 1381 and Jack Cade's Rebellion of 1540. A close examination of the play in comparison with its major sources reveals that Shakespeare departs from or expands upon the chroniclers' materials in such a way as to show a clearer, more pronounced distinction between the gentry and the baseborn in society and also to emphasize the absolute importance of maintaining the established ranks of blood in society. He is thereby the employer of conflict, an aesthetic advantage, between the conscious gentlehood of characters like Iden and *faisandé* upstarts like Cade.

It is important at the outset to note that Cade is a Kentishman, that the Kentish were spirited as a people, that they had no villeinage in their background, and that by their Germanic inheritance they were accustomed to electing (and deposing) their kings. Shakespeare's Cade is a Kentishman *manqué* whereas his Earl of Kent in *King Lear* is the apotheosis of the traditional Kentishman. Shakespeare, who never in this play mentions the Angles, Saxons, or Jutes, but who knew of the "ancient freedom" (IV.viii.27) of the Jutes, draws on descriptions of the Kentish from Julius Caesar and thereby melds Celt and Jute. Thus Lord Say in *2 Henry VI* describes them as people possessing noble qualities:

> *Kent, in the Commentaries Caesar writ,*
> *Is term'd the civill'st place of all this isle:*
> *Sweet is the country, because full of riches,*
> *The people liberal, valiant, active, wealthy.* (IV.vii.60–63)

The Kentish are also similarly praised by the Duke of York in *3 Henry VI*: "In them [Kentishmen] I trust, for they are soldiers, / Witty, courteous, liberal, full of spirit" (I.ii.42–43). Shakespeare in *King Lear* uses a type-name, Kent, as a prototype of the Kentish, designating the faithful yet spirited man who accompanies King Lear. The Kentish were thought free mainly because they fiercely resisted William the Conqueror at Swanscomb. William, impressed by their resistance, guaranteed that there should be no villeinage in Kent, thereby rendering Kent different from all other English counties. In *2 Henry VI*, however, Shakespeare, who (I repeat) knew the tradition of "ancient freedom," goes ahead with efforts, ignoring his sources, to make Cade a baseborn villein, albeit a high-spirited one, because he wants to show what happens in Elizabethan society if the legitimate blood-order yields its leadership to villeinous blood. *2 Henry VI* is basically a drama of rebellion, the dramatic procedure being that of York's usurpation, and there is a certain difference between Cade's rebellion and that of York. Even though York is shown as a very ambitious man, it is true that he has the right qualities to be a king and, more important, he has royal blood: "I am far better born than is the King: / More like a king, more kingly in my thoughts" (V.i.28–29). But Cade's rebellion is simply "the impious spectacle of the proper order reversed," a result of the baseborns' uprising. Attempting, as this essay does, to set Cade's villeiny in high relief is not to overlook the poet's skepticism evinced toward all, base and noble alike, who strive for power. But Shakespeare attributes Cade's sottishness to villeinous blood alone. Cade, therefore, is *sui generis*, a dreadful phoenix portentous of the commoners' total ineptitude for political rule, whom

Shakespeare goes out of his way to vilify, denigrate, degrade, and debase.

For Jack Cade in *2 Henry VI* we do not have any background information in the chronicles. The chronicles simply imply that since Cade and his followers are the Kentish, they are *ipso facto* not base by virtue of the Swanscomb agreement as codified in English law. In Act 4, Scene 2, however, Shakespeare invents Cade's father, mother, and wife, strongly emphasizing Cade's base blood in spite of his own pretensions to gentility. In the play his family members have vocations which fit the base, and Cade himself is a "clothier":

> *Cade.* *My father was a Mortimer -*
>
> *Dick.* *[Aside.] He was an honest*[15] *man, and a good bricklayer.*
>
> *Cade.* *My mother a Plantagenet -*
>
> *Dick.* *[Aside.] I knew her well, she was a midwife.*
>
> *Cade.* *My wife descended of the Lacies -*
>
> *Dick.* *[Aside.] She was indeed a pedlar's daughter, and sold many laces.*
>
> *Smith.* *[Aside.] But now of late, not able to travel with her furr'd pack, she washes bucks here at home.*
>
> *Cade.* *Therefore am I of an honorable house.*
>
> *Dick.* *[Aside.] Ay, by my faith, the field is honorable, and there was he born, under a hedge; for his father had never a house but the cage.* (IV.ii.39–52)

In Shakespeare's plays some gentles have been brought up by foster parents who are baseborn, but since those who raised Cade are his real parents, Cade must be a baseborn. Belarius in *Cymbeline* states, "Cowards father cowards and base things sire base" (IV.ii.26). Thus, Shakespeare with his blood-conscious attitude provides Cade, even bespatters him, with base blood.

Gentles in Shakespeare's plays seldom marry the baseborn because Shakespeare (as judged by his writings and by his approval of Dr. Hall, a gentleman, as his daughter's husband) and many of his contemporaries believed that high blood became degenerate or stained by marriage with people of low blood. Lawrence Stone observes that marriages of nobles with rich merchants' daughters, especially widows, were sometimes undertaken in the period 1558–1641 "after the successful production of a male heir by a first well-born wife had ensured that the hereditary line was saved from any taint of contaminated blood."[16] Moreover, gentles condemned members of their class and family who derogated the family blood by acquiring base characteristics or affecting the society of the baseborn. High blood, like blood of low degree, was not entirely a possession singular to an individual. It was shared especially with a wife because semen was in Renaissance times regarded as white-hot blood. In *2 Henry VI,* by depicting Cade as being derived from a baseborn father (like Shakespeare) and a baseborn mother (unlike Shakespeare), the poet doubly emphasizes Cade's base blood.

As the dialogues of Dick and Smith show, Shakespeare's Cade, unlike the Cade of the chronicles, is a beggar, a profession most repellent to gentles not degenerate or *déclassé.* Orlando in *As You Like It* does not beg food from Duke Senior and his men when he is starving: he threatens them with his sword. Cymbeline in stigmatizing Posthumus wrongly equates his poverty with his blood, importing modern economic notions of class into a society based on blood typing alone: "Thou [Imogen] took'st a beggar, wouldst have made my throne / A seat for baseness" (*Cymbeline* I.i.141–42). Moreover, throughout Shakespeare's plays *beggar* is employed to indicate the lowest class (e.g., *Measure for Measure* III.ii.83, *Much Ado About Nothing* III.iv.30, *The Winter's Tale* II.i.87, *Pericles* II.i.90), and this word is even used as a curse (e.g., *Richard III* I.ii.42). Cade was also a thief, a stealer of sheep, a crime quite unsuitable, unlike treason, for example, to persons of noble birth. Shakespeare generally adheres to the principles of the gentleman explained in Henry Peacham's *The Compleat Gentleman* in which Peacham states: "Nobilitie is the Honour of blood in a Race or Linage ... and since Honour is the reward of Vertue and glorious action onely, Vice and Basenesse must not expect her favours."[17] By making Cade a thief, something of which there is no mention in the chronicles, and by thus making him subject to hanging,[18] Shakespeare emphasizes Cade's lack of gentle blood. Furthermore, Cade was "whipt three market-days together"; Shakespeare never lets the gentles, however criminal, suffer from such a humiliation.

Shakespeare's Cade is ignorant and irrational, whereas the Cade of the chronicles, as Wilson and Stirling point out, was a man of "pregnaunt wit"[19] and he was "sober in talke, wise in reasoning"[20] when he met the king's representatives. Since only high blood endowed its possessor with the ability to learn or to possess learning without labor, as with Henry V, Shakespeare would not allow his Cade to have "pregnaunt wit," "sober" talk, or "wise" reasoning. Surely his hereditary blood is to some degree responsible. Cade's stupidity, irrationality, and over-passioned temper might come, in part, from foods. Being base meant eating base foods, such as beef. In Elizabethan England's conservative medical theory, the difficulty with beef was that it made thick, black, and melancholy veinal blood. John Huarte in an often-reprinted work points out that beef along with some other foods makes its eaters "have strength like a bull: but withall be furious and of a beastly wit."[21] In the play Cade is ignorant, a hater of literate people as well as the gentle, whereas in Hall's chronicle Cade was advised by "techers" and "privye scholemasters": Shakespeare's Cade hangs the clerk of Chartham with his pen and inkhorn about his neck because he can write his name, and Cade calls Lord Say a traitor because Say can speak French. Shakespeare intensifies the chronicles of Hall and Holinshed in their account of Cade's extreme irrationality in describing the lopped heads of Lord Say and his son-in-law Cromer: Shakespeare's Cade says, "Let them kiss one another, for they lov'd well when they were alive. Now part them again, lest they consult about the giving up of some more towns in France" (IV.vii.130–33). Shakespeare, furthermore, makes the scene of the executions of Say and Cromer more vigorous and abominable than do the chronicles so as to draw attention to Cade's irrationality. In Scene 7 Cade kills Say mainly because Say has "most traitorously corrupted the youth of the realm in erecting a grammar school" and had men about him "that usually talk of a noun and a verb, and such abominable words as no Christian ear can endure to hear" (IV.vii.32–41). Cade insists that Say put the poor men in prison because they could not read. Christopher Hill, one must note, states that Cade's "hatred of education" can be justified because a third of those sentenced to death in Shakespeare's London "escaped by pleading benefit of clergy,"[22] meaning essentially the ability to read. But in the play Cade's hatred does not fall on the educated only, but extends to all aristocrats.

Shakespeare augments Cade's cruelty further than what is seen in the chronicles. In the play Cade and his followers kill people without pity and destroy parts of the city without thinking because the baseborn (on the basis of analogy with baseborns in other Shakespearean plays) have thick blood: their basic condition tends to inhibit feelings of pity and kindliness. In *Macbeth*, for instance, the degenerate Lady Macbeth shouts, "Come, you spirits ... Make thick my blood, / Stop up th' access and passage to remorse, / That no compunctious visitings of nature / Shake my fell purpose, nor keep peace between / Th' effect and [it]" (I.v.40–47). If one follows Paul Kocher's interpretation,[23] as distinct from the more usual view that Lady Macbeth is inviting demon possession, one finds the blood of Scotland's future queen to be very like that which Shakespeare silently imputes to Cade and his accomplices. The body's melancholy, supplied with fresh melancholy from the spleen, would flow toward the heart, blocking the will from conscience-directed action. Put in another way, the base lacked animal spirits, the highest of three kinds of spirits in the human body. They in effect were incapable of enforcing their own moral judgments.

We can see that Shakespeare in other ways distances himself from the relatively classless treatment of Cade he saw in the chronicles, his sources. Cade in Hall's chronicle controlled his followers well at first and had a good strategy against Stafford: Cade caused his own army to retreat in order to lure the king's army into a disadvantageous position. But Shakespeare makes Cade and his followers a rabble: Cade commands them, "then are we in order when we are most out of order. Come, march forward" (IV.ii.189–90). In both chronicles, as Wilson points out, when Cade came into Southwark knowing that the king had fled to the castle of Killingworth in Warwickshire, he prohibited "to all his retinue, murder, rape, and robborie; by which colour of well meaning he the more allured to him the harts of the common people."[24] Cade in the play never gives that kind of thoughtful (or politic) command.

Cade's plans for reformation also attest to his irrationality and his projecting onto the vulgar the vast vulgarity of his mind. Even though there is no sympathy or admiration for the rebels in the chronicles, their demands are recorded fairly. In Holinshed Cade sent forth a "humble supplication, affirming that his comming was not against his grace, but against such of his councellours, as were louers of themselues, and oppressors of the poore commonaltie; flatterers of the king, and enimies to his honor; suckers of his purse, and robbers of his subiects; parciall to their freends, and extreame to their enimies; thorough bribes corrupted, and for indifferencie dooing nothing."[25] But in Shakespeare's play the content of Cade's supplication is not introduced, except the vow to kill Lord Say. Instead Cade has ridiculous plans of reform which show his half-baked populism: for example, "There shall be in England seven halfpenny loaves sold for a penny; the three-hoop'd pot shall have ten hoops, and I will make it felony to drink small beer" (IV.ii.65–68).

In Shakespeare's plays, moreover, the base usually deliver themselves in flat prose whereas gentles have figurative richness and musicality in language, and Cade is no exception. Even though Guiderius and Arviragus in *Cymbeline* have no courtly training or good education (although they have had long association with Belarius), they demonstrate ability to use both good rhetoric and verse, which is expected of persons of royal blood. Also in *The Winter's Tale* Perdita, who has been raised by a shepherd for sixteen years, uses refined language which, among other things, causes Polixenes and Camillo to infer the nobility of her blood. In *2 Henry VI* every gentle uses verse with eloquence in his dialogue, but Cade and his followers deliver themselves of flat, prosaic insults and incoherent prose. Cade's speech is unmemorable stylistically, and what can be remembered of it is its outrageousness.

Cade's unshapely appearance, quite different from the Cade in Hall who had a "goodely stature," also proves his baseness. High blood should be exhibited in good bodily contours and a handsome, perhaps even a distinguished, face; in Elizabethan language, a gentle should be well favored and well featured.[26] In *Pericles* Marina tells Leonine, "You are well favored, and your looks foreshow / You have a gentle heart" (IV.i.85–86): Marina is by implication complementing Leonine on his unstained blood derived from gentle and therefore comely parents. Theseus in *A Midsummer Night's Dream* says to Hermia that her father "composed your beauties" (I.i.48). In *2 Henry VI* the physical characteristics of the baseborn Cade are contrasted with those of the gentle Alexander Iden: Iden says, "See if thou canst outface me with thy looks. / Set limb to limb, and thou art far the lesser; / Thy hand is but a finger to my fist, / Thy leg a stick compared with this truncheon; / My foot shall fight with all the strength thou hast, / And if mine arm be heaved in the air, / Thy grave is digg'd already in the earth" (IV.x.46–52). Here Shakespeare stresses Cade's base blood through depicting him as a man of mean and even despicable appearance. Shakespeare's plays and many other writings of the period show a Neo-Platonic association of a fine appearance with virtue, and an ugly face and deformed body with vice.

Shakespeare's Cade also has bad breath—sure evidence of base blood, being the effluvia thereof. The chronicles are silent on this favorite Shakespearean way of diminishing the non-armigerous. In the Shakespearean plays the gentle have fragrant breath or, much more usually, no smell either of breath or body: in *Cymbeline* Arviragus says that the supposedly dead Fidele, the disguised Imogen, had good breath—"no, nor / The leaf of eglantine, whom not to slander, / Outsweet'ned not thy breath" (IV.ii.222–24). On the other hand, the base usually have bad breath: Brutus in *Coriolanus* deprecates seeking the approval of the base as begging "their stinking breaths" (I.i.61); Casca in *Julius Caesar* says that "such a deal of stinking breath" of the "rabblement" choked Caesar and made him swoon, and Casca "durst not laugh, for fear of opening [his] lips and receiving the bad air" (I.ii.244–50). It is highly significant that in *2 Henry VI* Cade's followers, themselves presumably vessels of stinking melancholy,[27] are offended by their leader's halitosis: "his breath stinks with eating toasted cheese" (IV.vii.12).[28] Cade is *faisandé* indeed.

The beer that Cade has been drinking attests to his base blood both by way of his preference and also *ex opere operato.* The chronicles of course say nothing of Cade's drinking habits. Shakespeare specifies wine as the gentle beverage and beers and ales, especially small beer, as beverages for the baseborn. Wine, which is called "the subtle blood o' the grape" in *Timon of Athens,* was known as an enhancer of the temperature in noble blood—indeed it was a temporary substitute (as with Falstaff) for blood; beer and ale, according to Berkeley, are "usually reckoned somewhat rebellious to assimilation into the blood and therefore required expense of irreplaceable 'radical' as well as fluent heat."[29] To lose heat meant to lose blood, which meant in turn the diminution of physical power, the power to enforce moral judgments, and shortening of life. So, compared with the gentle, the base, who also lost blood through sweat, tears, and hard work, begot their children at lower body temperatures with semen derived from blood of diminished heat, quantity, and quality. In Shakespearean times, wine was hard for the base to procure because of its high price. Thus Cade boasts that when he is king of England, "I will make it felony to drink small beer" (IV.ii.67–68), and "I charge and command that, of the city's cost, the pissing-conduit run nothing but claret wine this first year of our reign" (IV.vi.2–4). All of this is Shakespeare's lively invention.

Shakespeare's blood consciousness is also shown when he ridicules Cade's aspiration to climb to a higher social status. According to the chronicles, Cade at first rebelled against the gentle in order to "cast off the yoke of bondage and recouer libertie [of the commons],"[30] but, in the play, it seems that Cade gathers the base from the beginning of the episode only to actuate his desire to be a gentleman. More importantly, he desires to be king of England and equally to keep his base followers in their base condition:

> *Cade.* *... and in Cheapside shall my palfrey go to grass; and when I am king, as king I will be -*
>
> *All.* *God save your Majesty!*
>
> *Cade.* *I thank you, good people - there shall be no money; all shall eat and drink on my score, and I will apparel them all in one livery, that they may agree like brothers, and worship me their lord.* (IV.ii.68–75)

Shakespeare continuously and egregiously ridicules Cade's attempt to be a gentleman: when Cade meets Sir Humphrey Stafford and his brother, he knights himself by kneeling and saying, "Rise up, Sir John Mortimer" (IV.ii.120). In the chronicles, Cade is depicted as having some aspirations toward elevating his social class; Shakespeare greatly intensifies these aspirations. For example, in the chronicles he wears Sir Humphrey's armour. Hall's chronicle also states that Cade killed those who knew his base lineage. In the play Cade orders his followers to call him Lord Mortimer and kills a soldier who calls him Jack Cade. Shakespeare's Cade is in other ways extremely hubristic: in London after ordering the killing of Say, he orders that "The proudest peer in the realm shall not wear a head on his shoulders, unless he pay me tribute; there shall not a maid be married, but she shall pay to me her maidenhead, ere they have it" (IV.vii.114–17). In the first requirement Cade is inordinate beyond the practice of absolute kings, and in the second requirement he arrogates to himself, presumably for all England, the traditional *droit du seigneur* of a medieval lord of a manor. He also irrationally orders the burning of "all the records of the realm, my mouth shall be the parliament of England" (IV.vii.14–15), a piece of outrageousness that elicits scoffing even from his own followers.

Shakespeare derides not only Cade but also his base followers: they are also ignorant, vulgar, destructive, and politically unstable. They hate literate people: they want to kill all lawyers, as Holinshed states, and, going beyond the chronicle, Shakespeare's mobs heap abuses on the Clerk of Chartham simply because he can write his name, calling him "a villain and a traitor" (IV.ii.108). Shakespeare also distinguishes fighting of the baseborn from that of the armigerous. In Act 4, Scene 3, Dick fights like a butcher, his vocation: Cade praises Dick, saying "They fell before thee like sheep and oxen, and thou behavedst thyself as if thou hadst been in thine own slaughter-house" (IV.iii.3–5). As a reward for meritorious services in war, Cade gives him "a license to kill for a hundred lacking one" during Lent and Dick desires "no more" (IV.iii.7–9). Furthermore, Cade's followers are, as Stirling points out, politically unstable and mutable like the commoners in *Julius Caesar* and *Coriolanus*—a direct consequence of their low blood quality: they are "so lightly blown to and fro" between Cade and Clifford.

Throughout the Cade episode, Shakespeare depicts Cade and his baseborn followers as being notably debased human beings. The episode intensifies a dominant view of Shakespeare's contemporaries: heredity very largely determines human individuals. While the gentle have a good appearance, the ability to learn quickly, a wider range of emotions, the strength to enforce their own moral judgments, and rational control of the passions, the ugly, deluded, uninformed, impulsive, poor baseborns do not enjoy these advantages. Moreover, whereas the gentle are eventually rewarded, the base, as the Cade episode shows, are always punished if they aspire beyond their lowly status; Shakespeare obviously believes that base blood does not deserve such recompense. In the Cade scenes Shakespeare's modification of source materials is evidence of this belief in blood and class-consciousness, demonstrating that Shakespeare is the arch-conservative, the most strict believer in the merits of the gentle and the demerits of the base. The poet in this important regard is unlike the Chaucer of "The Clerkes Tale," the Marlowe of *Doctor Faustus*, the Dekker of *The Shoemaker's Holiday,* the Webster of *The Duchess of Malfi,* and the Milton of *Samson Agonistes*; he is unlike Mary of the Magnificat; and his plots show nothing like the Dick Whittington story. Although the idea of blood gentility once had great currency—even *La Marseillaise* refers to "the stained blood" of enemies of the Revolution—the idea of a blood-based aristocracy is now so *passé* that David Cannadine does not even mention its disappearance in his work on the British aristocracy.[31] As to the truth of Shakespeare's general position and that of other Tudor conservatives, it is perhaps wisest to remember Goethe's dying remark: "Mehr Licht!"

Notes

1 Charles Hobdays, "Clouted Shoon and Leather Aprons: Shakespeare and the Egalitarian Tradition," *Renaissance and Modern Studies* 23 (1979): 63–78.

2 Michael Hattaway, "Rebellion, Class Consciousness, and Shakespeare's *2 Henry VI,*" *Cahiers Elisabethains: Etudes sur la Pre-Renaissance et la Renaissance Anglaises* 33 (1988): 13–22.

3 J. P. Brockbank, "The Frame of Disorder - 'Henry VI'" in *Early Shakespeare,* ed. John Russell Brown and Bernard Harris, Stratford-upon-Avon Studies 3 (New York, 1961) 73–110.

4 M. M. Reese, *The Cease of Majesty: A Study of Shakespeare's History Plays* (New York, 1961) esp. 126.

5 Brents Stirling, *The Populace in Shakespeare* (New York, 1949).

6 Richard Wilson, "'A Mingled Yarn': Shakespeare and the Cloth Workers," *Literature and History* 12.2 (1986): 164–80.

7 Karl Brunner, "Middle Class Attitudes in Shakespeare's Histories," *Shakespeare Survey* (1953): 36–38.

8 Shakespeare wrote the Henry VI plays before he became a gentleman in 1596.

9 Stirling 24.

10 Wilson 167.

11 Brunner 37.

12 Irving Ribner and George Lyman Kittredge, eds. *The Complete Works of Shakespeare* (New York, 1971) 551.

13 Thomas Walkington, *The Optick Glasse of Humors* (London, 1639) 1.

14 Andrew Boorde, *A Compendyous Regyment or a Dyetary of Helth* (1542) EETS Extra Series 10 (London, 1870) 2.

15 "Honest" is often a class term with a derogatory subaudition.

16 Lawrence Stone, *The Crisis of the Aristocracy 1558–1641* (Oxford, 1965) 628.

17 Henry Peacham, *The Compleat Gentleman* (London, 1622) 2–3.

18 He would not be decapitated, a fate reserved for the gentle.

19 Edward Hall, "The Union of The Two Noble and Illustre Famelies of Lancastre and Yorke" in *Narrative and Dramatic Sources of Shakespeare*, ed. Geoffrey Bullough, 8 vols. (London, 1966) 3:113.

20 Raphael Holinshed, *Holinshed's Chronicle as Used in Shakespeare's Plays* (London, 1943) 116.

21 John Huarte, Examen des Ingenios. *The Examination of Mens Wits*, trans. R. Carew (London, 1594) 305.

22 Christopher Hill, "The Many-Headed Monster in Late Tudor and Early Stuart Political Thinking" in *From the Renaissance to the Counter-Reformation: Essays in Honor of Garrett Mattingly*, ed. Charles H. Carter (New York, 1965), 303.

23 Paul Kocher, "Lady Macbeth and the Doctor," *Shakespeare Quarterly* 5 (1954): 34–49.

24 Holinshed 116.

25 Holinshed 116.

26 There is a notable anomaly in Shakespeare's plays: Talbot in *I Henry VI* is described as a "writhled shrimp" (II.iii.23) yet is of such high personal honor as to be an arbiter of the membership of the Order of the Garter.

27 John Huarte states that some foods, such as beef, manzo, red grain bread, cheese, olives, and vinegar make gross sperm and "faultie temperature." Stirling devotes several pages to what he felicitously calls "proletarian stench" especially characteristic of Shakespeare's vulgar, but he never mentions the Galenic melancholy of the base as an ineradicable source of body odor, by analogy emitted by Cade and his followers. Lawrance Babb in *The Elizabethan Malady* (East Lansing: Michigan State College Press, 1951) and Bridget Lyons in *Voices of Melancholy: Studies in Literary Treatments of Melancholy in Renaissance England* (New York: Barnes & Noble, 1971), discussing Elizabethan melancholy, likewise omit this important cause of body odor.

28 Toasted cheese along with garlic and onions (*A Midsummer Night's Dream*, IV.ii.42–43; *Coriolanus*, IV.vii.98) was regarded as diet which helps to make the base's breath bad.

29 Berkeley 54.

[30] Holinshed 115.

[31] David Cannadine, *The Decline and Fall of the British Aristocracy* (New Haven, 1990).

The Stained Blood of Rape: Elizabethan Medical Thought and Shakespeare's *Lucrece*

by Randy Phillis

Critics who have studied William Shakespeare's *The Rape of Lucrece* (1594) often focus their attention on the justification of Lucrece's decision to commit suicide, carrying on an ancient debate begun by St. Augustine long before Shakespeare adopted the story for use in his poem.[1] Less concerned with the morality of her act than earlier critics were, modern critics instead have examined various other justifications for her behavior, making arguments based upon either Lucrece's personal reasons or her obligations to both her family and society, or else they offer various metaphorical explanations. However, for the most part, critics have ignored the fact that Shakespeare provides a literal, medical justification, in addition to others, for Lucrece's act of suicide.

Much recent criticism has focused upon Shakespeare as a Tudor conservative, a writer very conscious of the quality of his characters' blood.[2] This criticism suggests that Shakespeare was well aware of Elizabethan medical thought regarding the importance of blood quality and that he employs this belief as a motif in his works so consistently that we may justify characters' behaviors by looking at the condition of their blood. Applied to Lucrece, the concept is a simple one: she commits suicide because her blood has literally been stained by Tarquin's rape. He has left an indelible blot in her blood, and the only way for her to escape the influence of this bad blood is death. Thus Shakespeare provides a neutral, nonmoral, scientific explanation for Lucrece's need to die.

Modern critics have come up with a variety of other justifications for Lucrece's suicide, including the idea that Shakespeare was trying to present a moral lesson. Louis R. Zocca believes that much Elizabethan poetry, including *The Rape of Lucrece,* was used "to drive home the lessons of chastity and virtue" and to show the abhorrence of "lust and concupiscence."[3] However, Heather Dubrow disagrees, writing that such complaint poems as *Lucrece* were intended to "move their readers" rather than instruct them.[4]

Other critics view Lucrece's decision as a moral or psychological triumph. Saad El-Gabalawy, Ian Donaldson, Laura G. Bromley, and Dubrow all believe that Lucrece is engaged in a "moral struggle."[5] These critics believe that for Lucrece there is no "middle-ground"[6] between honorable and dishonorable conditions, and that she finds this predicament "more intolerable than death itself."[7] Bromley also believes that the suicide is Lucrece's psychological attempt to regain "the balance of herself" and "restore order" to her divided nature.[8] Donaldson and Katherine Maus see Lucrece as trying to restore herself, psychologically, to the condition she was in before the rape.[9] While some critics simply assert that Lucrece is overly concerned with her personal reputation and honor, others believe that the idea of honor goes beyond Lucrece herself, extending to the reputation of the Collatine family and even to the very social order itself. El-Galabaway points out that "the stain [the emblem of dishonor from the rape] inevitably extends to her husband, to her family, to society at large."[10] Her suicide is then justified by "citing obligations to her family."[11] This preservation of family as an institution extends naturally to a preservation of society as a whole. Lucrece "upholds the social order" by dying for "the sake of marriage as an institution"[12] as well as by forcing the men into the action necessary to remedy the currently corrupt political system.[13] Lucrece's suicide can also be justified according to ancient Roman religious views.[14]

Two critics, citing an abundance of primary evidence in the poem dealing with the stain, provide reasonable metaphorical explanations for Lucrece's suicide. Maus sees the "tragic dividedness"[15] of Lucrece's blood as a metaphor signaling "inescapable" danger for Lucrece.[16]

Kahn, in the oft-cited work "The Rape in Shakespeare's *Lucrece*," identifies Lucrece's stain as "the central metaphor" in the poem.[17] Kahn systematically shows how Lucrece is "morally 'stained' and sexually 'tainted.'"[18] Kahn believes that Lucrece's "moral and social purpose as a wife have been destroyed," a situation which results in Lucrece's believing that a physical change has taken place in her blood which is "symbolized for the reader when her blood ... divides into two streams upon her death."[19] This is an understandable explanation for the two streams of blood evident after Lucrece's suicide, but I believe that the streams were not intended to be symbolic. They are, in fact, real—literally two streams of blood, one still pure and one actually physically stained by the rape.

While I cannot say that all these critics are necessarily incorrect in their views, they have all overlooked the physical evidence that Lucrece's blood is literally tainted by the rape and that the predicament is a physical one for which, given Lucrece's honor, death is the only solution. This view, commonly held among Elizabethans, makes moral and political arguments secondary, and it is readily supported by a brief examination of Elizabethan medical thought. One critic, Bickford Sylvester, noted this view as a tangential possibility in "Natural Mutability and Human Responsibility: Form in Shakespeare's *Lucrece*," though his thesis is that Lucrece's act was essentially a political one. Unfortunately, Sylvester does not pursue the medical line of thought and consequently offers little evidence (none of which is primary) for the very point which I believe explains the suicide quite clearly.

A brief survey of Elizabethan medical texts shows that Shakespeare's audience might well have considered Lucrece's stain as literal, something which would explain—physiologically—the necessity of Lucrece's death. Once literally contaminated by Tarquin's blood, Lucrece was trapped; moral and social issues became secondary. After the rape, Lucrece can never be the same physically, and indeed all her "issue" will be "blurr'd with nameless bastardy" (522). The stain affects her physically in such a way that she will never be capable of bearing a pure Collatine line—her primary responsibility as the wife of a patrician.

Elizabethan physiological thought, based heavily on Galen and Aristotle, was founded upon the idea that everything in the human body—including flesh, bone, and hair—was blood in a concocted or refined form.[20] After going through several stages, blood was changed into everything necessary for life.[21] Semen was no exception, as according to Aristotle it was quite simply "blood as it becomes concocted."[22] Jacques Ferrand, writing in 1640, explains the point most succinctly: "the seed is nothing else but Blood, made White by the Naturall Heat."[23] Tarquin's semen, then, is corrupt like his blood; in fact, it is simply his blood in a slightly different manifestation. Once the semen is deposited into the uterus it cools down, once again becoming blood which then mixes readily with the woman's.[24] This depositing of seed thus affects the woman's blood quality, either ennobling her or debasing her, depending upon the relative quality of each party's blood. A similar situation occurs in *Hamlet*, according to David Shelley Berkeley in "Claudius the Villein King of Denmark." Berkeley writes that one reason Hamlet is disturbed by Queen Gertrude's marriage to Claudius is that because she is having sex with Claudius, Gertrude is becoming "villeinized with Claudius' thick, cold, sluggish, ill tasting, darkish blood (and therefore acquiring all his vicious and ugly qualities)."[25] Once Tarquin's semen has reverted to blood inside Lucrece and has mixed with her own, she literally "bears the load of lust he left behind" (734) in the form of tainted blood.

Distribution of the tainted blood throughout Lucrece's system is rapid, following a direct route through her whole body. Because the "blood-vessels join on to the uterus as though they were roots,"[26] the tainted blood travels directly to the heart and through the whole body.[27] Lucrece is doomed. And not only is she in immediate physical trouble, but she stands to lose control of her mental faculties as well; as Ferrand points out, bad semen has the ability to become "corrupt within the body and so by sending up diverse noysome vapours to the Braine by the Back-bone & other passages … disturb and hinder the operations of its chiefist Faculties."[28] Lucrece's "fountain" is made muddy in a way "that cannot be amended" (577–78) because her ocean of blood is literally stained (654–55). She is trapped in a "polluted prison" (1726), and Collatine's "stock" (1063) is permanently ruined. Collatine himself, in fact, runs the risk of being poisoned by Lucrece's "attaint" should he have sex with her now that she has been raped. Lucrece says, "I will not poison thee with my attaint" (1072). The word *attaint* has been used earlier in the poem in conjunction with "shame" and "Disgrace" (825; 827) where it means "dishonor."

But attaint also carries the meanings "to infect with corruption, poison"[29] and, more significantly, "corruption of blood."[30] It is clear that Shakespeare knew these meanings, as he had used the word in that sense in *Venus and Adonis*:

As burning fevers, agues pale and faint
Life-poisoning pestilence, and frenzies wood,
The marrow-eating sickness, whose attaint
Disorder breeds by heating of the blood. (739–42)

That Lucrece's blood is literally stained by being mixed with Tarquin's is proved upon her death. Her once pure blood (440) has now become, like Priam's in the picture Lucrece looks upon, black:

Her blue blood chang'd to black in every vein , ...
Show'd life imprison'd in a body dead (1454;1456).

When the knife is removed from Lucrece's breast, the blood literally separates, just as we might expect:

And bubbling from her breast, it doth divide
In two slow rivers, that the crimson blood
Circles her body in on every side, ...
Some of her blood still pure and red remain'd,
And some look'd black, and that false Tarquin stain'd. (1737–39; 1742–43)

The color and consistency of Lucrece's blood, both the pure and the tainted, is significant. It is physical proof of what Elizabethans understood to be a fact about blood quality. Lucrece's pure, original blood "untainted still doth red abide"(1749) while the "Corrupted blood some watery token shows"(1748) and is "black" (1745). According to Alexander Read in *The Chirurgical Lectures of Tumors and Ulcers*, Elizabethans believed that "Good blood is discerned by colour, taste and consistence: In colour it is red ... and consistence meane, between thick and thin ...."[31] Robert Burton also explains that good blood is a "temperate red humour."[32] Bad blood, like that deposited in Lucrece by Tarquin, is "darker."[33] "Unnaturall" conditions are at work if the blood appears "blacke ... unequall, of diverse substances"[34] and this condition, when the blood is "thick" or "thin" and "blacker,"[35] can prove fatal in itself.[36] Lucrece is thus physiologically justified when she arrives at the solution to her problem: "The remedy indeed to do me good / Is to let forth my foul defiled blood" (1028–29).

Even if Lucrece were to choose life, she could no longer perform her primary function of bearing Collatine's children and continuing his bloodlines. Her "issue" would indeed be "blurr'd with nameless bastardy" (522), though it may be difficult for modern readers to understand how one sexual act—even if no pregnancy resulted—could so permanently destroy bloodlines. The fact is that Lucrece's blood is now permanently impure: No matter how much the impurity may be diluted by time or other factors, that impurity will invariably manifest itself in children subsequently born. According to Francis Markham, pure, untainted blood was absolutely necessary for generations in order to make and keep a family great.[37] No blemish, no matter how small or how distantly attained in the past, could ever be remedied. Once raped, Lucrece was destined to bring "forth no absolute, but an imperfect generation."[38] There was no middle-ground in such a situation, as blood was considered "either *Legitima*, or *Notha*, either truly bred, or else bastardly."[39] In fact, according to John Huarte, once the bloodline was stained, "the husbands children [would] resemble the adulterer though they be not his."[40] This was due to the pre-natal influence of the woman's thoughts about the adulterer; to the quality of the woman's blood, tainted by the adulterer, which would be passed directly on to the children through her seed and to the fetus's nourishment in the corrupted womb.[41] Aristotle illustrates how such a tainting can influence subsequent generations, even though the stain may be recessive in the first generation: "There was at Elis a woman who had intercourse with a blackamoor; her daughter was not a black, but her daughter's son was."[42]

Any children feeding on the mother's blood in the womb and on her milk after birth would necessarily be affected by the condition of her blood, a point made by John Makluire in *The Buckler of Bodilie Healthe.*[43] And, as mentioned above, Collatine himself appears to run the risk of debasement simply by having sex with his now-stained wife.

Thus Lucrece's predicament is an inescapably physical one, and her decision to commit suicide need not be based on moral or honorable considerations at all. Her decision does, of course, carry moral, reputational, and political implications, but these I suggest become secondary when it is understood that she has become—quite literally and physically—a different person from the woman she was before the rape, and that the original person cannot be restored.

Tragically, she is no longer a Collatine in the pure sense she once was, and there is nothing she or anyone else can do about it.

Notes

1 Ian Donaldson's *The Rapes of Lucretia* (New York: Oxford UP, 1982) traces the story of Lucrece and its adaptations from its origin in ancient times, and it discusses the critical debates which have accompanied the story since Augustine. A briefer, though very informative, survey of the classical critical debates is presented in Saad El-Gabalaway's "The Ethical Question of Lucrece: A Case of Rape," *Mosaic* 12 (1979):75–86.

2 David Shelley Berkeley's *Blood Will Tell in Shakespeare's Plays* (Lubbock: Texas Tech UP, 1984) gives a convincing account of how Shakespeare's characters may be typed by genetics and physiology, especially in regard to blood quality.

3 Louis R. Zocca, *Elizabethan Narrative Poetry* (New Brunswick: Rutgers UP, 1950) 53.

4 Heather Dubrow, "A Mirror for Complaints: Shakespeare's *Lucrece* and Generic Tradition," *Renaissance Genres: Essays of Theory, History, and Interpretation*, ed. Barbara Kiefer Lewalski (Cambridge: Harvard UP, 1986) 401.

5 Laura G. Bromley, "Lucrece's Re-Creation," *Shakespeare Quarterly* 34 (1983): 201.

6 Dubrow 409.

7 Donaldson 22.

8 Bromley 202.

9 Katharine Eisaman Maus, "Taking Tropes Seriously: Language and Violence in Shakespeare's *Rape of Lucrece*," *Shakespeare Quarterly* 37 (1986): 70.

10 El-Galabaway 81.

11 Dubrow 416.

12 Coppelia Kahn, "The Rape in Shakespeare's *Lucrece*," *Shakespeare Studies* 9 (1976): 49.

13 Bickford Sylvester, "Natural Mutability and Human Responsibility: Form in Shakespeare's *Lucrece*," *College English* 26 (1965): 510.

14 Donaldson 24–5.

15 Maus 72.

16 Maus 82.

17 Kahn 47.

18 Kahn 46.

19 Kahn 64.

20 Galen, *On the Natural Faculties*, trans. Arthur John Brock (Cambridge: Harvard UP, 1963) 131.

21 Galen 131.

22 Aristotle, *Generation of Animals*, trans. A.L. Peck (Cambridge: Harvard UP, 1963): 91.

23 Jacques Ferrand, *Erotomania or A Treatise Discoursing of the Essence, Causes, Symptoms, Prognostics, and Cure of Love or Erotic Melancholy* (London, 1640) 261.

24 Galen 133.

25 David Shelley Berkeley, "Claudius the Villein King of Denmark," *Hamlet Studies* 11 (1989): 9.

26 Aristotle 197.

27 Aristotle 471.

28 Ferrand 261.

29 *Oxford English Dictionary*, 10. This usage dates from 1580 to 1856.

30 *Oxford English Dictionary*, 6. 1340–1844.

31 Alexander Read, *The Chirurgicall Lectures of Tumors and Ulcers* (London, 1635) 45.

32 Robert Burton, *The Anatomy of Melancholy*, ed. Floyd Dell and Paul Jordon-Smith (New York: Tudor, 1948) 128.

33 Galen 207.

34 Read 55.

35 Galen 169.

36 Galen 205.

37 Francis Markham, *The Booke of Honour or Five Decades of Epistles of Honour* (London, 1625) 46.

38 Markham 47.

39 Read 45.

40 John Huarte, *Examin de Ingenios* (London, 1594) 309.

41 Huarte 319–21.

42 Aristotle 55.

43 John Makluire, *The Buckler of Bodilie Healthe* (London, 1630) 116. Mother's milk was also considered to be concocted blood and held the same power to transfer the generator's qualities as male semen. See also Aristotle 193, 473.

"Vengeance Rot You All!" Blood-Oriented Revengers in *Titus Andronicus*

by Myung-soo Hur

Titus Andronicus, Shakespeare's first attempt at tragedy, portrays major characters as revengers motivated by their staunch and spontaneous blood-consciousness. The poet's dramaturgical power in this play arises from his creation of vigorous conflicts between the antipodes of a revengeful gentry and a spiteful peasantry. Shakespeare, according to Ralph Berry, sees *revenge* as a "recurring human motive," for each of his plays exhibits at least one instance, sometimes many, of revenge or revenger except for *Love's Labor's Lost* (51). The *American College Dictionary* defines *revenge* as "retaliation for injuries or wrongs," as distinguished from *retribution,* which "suggests just or deserved punishment, often without personal motives." In the Renaissance, the word *revenge* had not only the same negative meaning as the modern one but also a more extended meaning, one that is nearly equivalent to today's *retribution* (Broude 39). The word indicated either retribution effected directly by an individual or family, or "public vengeance" executed by magistrates ("common revengers"), or "vengeance of the Lord" (41). The wide-sweeping entries of *The Oxford English Dictionary* include a good sense of the word as illustrated in the phrase "in revenge of," meaning "in recompense for." On the one hand, some religious minds might stress such passages as Romans 12.19—"Dearly beloued, auenge not your selues, but giue place vnto wrath: for it is written, Vengeace [sic] is mine: I wil repaye, saith the Lord."[1] On the other hand, other religious minds might remember such justified avengers as Samson

(Judges 16.28–30) or think of the passage in Numbers 35.19—"The reuenger of the blood him selfe shal slay the murtherer: when he meteth him, he shal slay him." With pejorative overtones, Francis Bacon in "Of Revenge" says, "Revenge is a kind of wild justice .... Certainly, in taking revenge, a man is but even with his enemy, but in passing it over, he is superior; for it is a prince's part to pardon." However, he cautiously adds, "The most tolerable of revenge is for those wrongs which there is no law to remedy" (15–16). Linda Anderson argues that Hieronymo in Thomas Kyd's *The Spanish Tragedy* as well as Shakespeare's own Hamlet are justified revengers because they wreak revenge on evil-doers where "there is no law to remedy," and that these justified revengers seem to have been the most popular in revenge tragedy (17). In short, the Elizabethans neither universally condemned nor universally condoned all acts of revenge.

The Shakespearean gentles, if not degenerate, appear to be justified revengers whose sense of honor drives them courageously to withstand and to rectify social injustice and evil. David S. Berkeley in *Blood Will Tell in Shakespeare's Plays* observes that high blood was practically synonymous with courage—"the *sine qua non* of gentility" (20)—and that the Shakespearean gentles have inherited the four classical virtues—"prudence, temperance, courage, and justice"—through the excellent blood of their parents (84). Whereas the baseborns bear injustice in a humiliating and cowardly manner, the gentles demonstrate their courage to fight against evil, restore good, and avenge wrongdoings inflicted on them and their families. In *Titus Andronicus,* it is a base act for gentlemen to bear dishonor without retribution, just as the new Roman Emperor Saturninus declares in a rhetorical question, "Be dishonored openly, / And basely put it up without revenge?" (I.i.433–34).[2] Hence Lucius wreaks vengeance on Saturninus when the Emperor stabs Titus, Lucius' father, justifying his revenge: "Can the son's eye behold his father bleed? / There's meed for meed, death for a deadly deed!" (V.iii.65). After his assassination of the Emperor, the "common voice" hails Lucius "Rome's royal emperor!" (140–41). Rather than enduring insults, gentlemen often seek revenge on the insulter or evildoer in a judicial combat or a duel, just as Bolingbroke and Mowbray try to do in a judicial combat before they are stopped by Richard II. By contrast, the baseborns like Aaron, though spiteful or revengeful, are always too cowardly or deceitful to fight *vis-à-vis* against the gentles in a duel or judicial combat or other gentlemanly encounter. They are promptly defeated if they venture to fight with the high-bloods. So they can but rely on tricky schemes—concrete expressions of Machiavellian policy and the art of dissembling. Most of the major characters in *Titus Andronicus* emerge as revengers for blood-based reasons: they dauntlessly confront such genetic issues as sibling rivalry, primogeniture, patriarchal authority, cross-class union, and racial and genetic bias. They know their status in the family and state; their unyielding minds to keep their positions or advance themselves to their desired or rightful status creates various conflicts, which in turn shape the structure of the whole play.

The characters in *Titus Andronicus* can be roughly divided into six groups according to their blood-quality: the four gentry groups—the ideal heroes (Marcus and Lucius), the innocent victims (Bassianus and Lavinia), degenerate gentles (Titus, Saturninus, Tamora, Demetrius, and Chiron), and other minor gentles (Young Lucius, Aemilius, a Roman Lord, etc.); and the two plebeian groups—the rustic clowns (the unnamed Clown and the nurse) and the evil peasantry exemplified by Aaron the Moor. Shakespeare uses his large cast to define characters by antithesis. The contention between the Andronici (the noble Romans) and Aaron (the evil Moor) offers a dramatic contrast in the play. The contrast between gentility (Lavinia) and degeneracy (Tamora) intensifies this conflict, inasmuch as vice in degenerate gentles is far worse than that in the plebeians: "Lilies that fester smell far worse than weeds" (Sonnet 94, line 14). The sibling rivalry between the degenerate Emperor Saturninus and his virtuous brother Bassianus serves as another thematic contrast. Titus' rigid sense of patriarchal authority and his belief in primogeniture function as the pivot of all the contrasts and oppositions which lead to the destruction of Rome: Aaron's voice resonates throughout the play—"Vengeance rot you all!" (V.i.58). And this voice, I imagine, will be echoed by another voice springing from the inner heart of Shakespeare—"Vengeance rot your bloods all!"

The whole structure of *Titus Andronicus* hinges on this authorial voice. The first act opens with the sibling rivalry between the sons of the late Emperor of Rome—Saturninus and Bassianus—over primogeniture, rightful inheritance, and legitimacy of the new emperorship. The first stage direction suits the mood of this conflict. As the Roman tribunes and senators gather on a higher level of the stage,

Saturninus enters with his followers at one door, and Bassianus and his followers at the other door, with drums and trumpets. Saturninus first claims the throne by right of primogeniture, reminding them that he is the late Emperor's "firstborn son" (I.i.5). Saturninus' claim is based on the idea that the eldest son, begotten by the father's youthful blood—healthy, rich, abundant—is more like his father than any of the other sons. Oliver in *As You Like It* is spiteful because his younger brother challenges his authority as the firstborn son; Orlando is discomfited because his brother abuses the authority of the firstborn. The conflicting tension between Saturninus and Bassianus appears to be much worse than that between Oliver and Orlando. Saturninus is conspicuously revengeful for his younger brother's challenge to an election. Bassianus, on the other hand, bases his own claim on his superior merits—"And suffer not dishonor to approach / The imperial seat, to virtue consecrate, / To justice, continence, and nobility" (13–15). This controversy is calmed temporarily by Marcus, who announces that his brother Titus has already been elected by the Roman citizens as the new Emperor for his victory over the Goths. The issue of primogeniture continues when Titus gives up the crown in favor of Saturninus—the late "emperor's eldest son" (225), and when he selects Alarbus, "the eldest son" of the Queen of the Goths (103) to be sacrificed for his dead sons. Tamora, Queen of the Goths, entreats Titus to spare her "firstborn son" by insisting that "Sweet mercy is nobility's true badge" (119). Titus rejects this appeal on the ground that Alarbus, representative of the Goths by reason of his being the firstborn son of the Queen, must be a suitable sacrifice for the Roman ritual to "appease their groaning shadows that are gone" (124).

The sibling rivalry concerning primogeniture grows more complex and dangerous when Saturninus vows to take Lavinia (Titus' only daughter) for his empress, although Lavinia has already been betrothed to Bassianus. As a token of gratitude for Titus' selection of him as emperor, Saturninus promises to promote his "name and honorable family" by requesting Lavinia as his bride and empress. But this seems to be only one of the reasons for his proposing to Lavinia. Shakespeare intimates that Saturninus also knows of his brother's alliance with Lavinia, for all the brothers of Lavinia and even her uncle Marcus apparently know of their betrothal. When Saturninus is about to leave with Lavinia, Bassianus seizes her and declares, "this maid is mine" (I.i.277). Marcus endorses Bassianus by saying, "Suum cuique [To each his own] is our Roman justice. / This prince in justice seizeth but his own" (281–82), and Titus' sons convey Lavinia away with Bassianus. Only Titus seems to be ignorant of the prior pledge of Lavinia to Bassianus, probably because of his long absence from Rome. If Titus knows of the betrothal and disregards it, then the couple must have been engaged without his approval. Whether or not Titus is informed of their betrothal, almost everyone else, including Saturninus, appears to know it. Why, then, does Saturninus claim his brother's "betrothed love" to be his own? To be sure, his outspoken reason is to honor the Andronici by marrying Lavinia. But he has another reason: he envies his brother for winning Lavinia, "Rome's royal mistress," because he too deeply falls in love with her—she is the "mistress of my heart" (242). Frequently, Elizabethan writers including Shakespeare, as Aubrey C. Kail observes, associate particular organs with particular emotions: spite with the spleen, love with the heart, and lust and courage with the liver (142). Saturninus' words—"mistress of my heart"—strongly suggest that he has already fallen in love with Lavinia, though his brother has won her own heart. Falling in love is characteristic of the sanguine gentles, as shown in the illustration in a medieval manuscript about the four humors of the sanguine man who embraces a fair gentlewoman.[3] The sibling rivalry is thus twofold—competing for Lavinia and for the emperorship. Saturninus may be entitled to neither of these, considering his reputation. Thanks to Titus' retirement and endorsement, however, he can earn one trophy—emperorship. The other trophy—Lavinia—is also valuable for him. When Titus nominates him to be Emperor, Saturninus' emotions are mixed with love for Lavinia and spite for Bassianus. One reason for Saturninus' claim to Lavinia, as Max H. James perceptively suggests, is to "spite Bassianus, who had challenged his right to be emperor" (31). He has already been revengeful since his brother won Lavinia, and he has become even more spiteful because of Bassianus' challenge. So the first action he takes as emperor is to claim Lavinia. Only if he could win Lavinia would Saturninus defeat his brother completely.

Shakespeare complicates the selection of the Emperor by fusing two different principles. The only legitimate way to be King in Elizabethan England is, as C. G. Thayer holds, "by fair sequence and succession" (80). But the Rome of Saturninus, like the Denmark of Hamlet, does not secure him emperorship on the basis of primogeniture.

It is an anachronism to conform the selection of Roman Emperor to an English practice. Titus and Saturninus are, in this sense, anachronistic to rely on the English principle. In Shakespeare, however, "fair sequence and succession" cannot always guarantee a good king, or even a tolerable one, as exemplified by Saturninus and Richard II. What matters to Shakespeare is the blood-quality of the ruler. Having high blood—"red, abundant, hot, thin, fast-flowing, and sweet-tasting"[4]—such ideal heroes as Henry V, Guiderius, and Lucius are morally innocent and can maintain peace and order in a kingdom and rule the people successfully. The degenerate rulers like Saturninus, Richard II, Richard III, and Macbeth, to name only a few, ruin not only their lives but also their countries.

Both Saturninus and Bassianus are royal by birth, being sons of the late Emperor. Although the elder brother suffers from a bad reputation, he still seems to be gentle of blood at the beginning of the play. He articulates his blood-based reason for his claim—"to let my father's honors live in me" as "his firstborn son"—and so he considers his younger brother's challenge with "indignity" (I.i.7,8). His pride in his high blood is aroused by his younger brother's challenge, which makes him even more vengeful and spiteful toward Bassianus, a complication adding to the love-triangle conflict. His high blood begins to deteriorate when he becomes enchanted at first sight of Tamora, even though he has just avowed his love for Lavinia. Just as the wicked Queen's beauty enchants Cymbeline, Tamora's erotic beauty bewitches Saturninus. Torn between two beauties (noble and erotic), he moans in an aside, "A goodly lady, trust me, of the hue / That I would choose, were I to choose anew" (262–63). Indeed, when Bassianus claims Lavinia, he unhesitatingly makes his "sudden choice" of Tamora as his wife and Empress (319).

In many ways, Tamora, Queen of the Goths and the mistress of Aaron the Moor, serves as the main cause of Emperor Saturninus' degeneration, which personifies the decline of Rome. Tamora's royal blood seems to have become degenerate by reason of her cross-class alliance with the evil plebeian—Aaron the Moor, an "incarnate devil" (V.i.40). Aaron, then, is the ultimate cause of Saturninus' degeneration and of the corruption of Rome, as well as of Tamora's tainted blood. The idea of degenerate blood derives from Aristotelian and Galenic conceptions. Galen in his *On the Natural Faculties* mentions the functions of blood as determining the formation of all parts of the human body. Thus Thomas Walkington in *The Optick Glasse of Humours* (1607) says, "In the elements consists the body, in the body the blood .... It is a nutriment for all and singular parts of what qualities soever" (58). According to Aristotle in *Generation of Animals*, "semen is concocted blood" (I.xix.91). So Thomas Cogan delineates the transformation of blood into semen in *The Haven of Health*:

> *After the third and last concoction: which is doone in everie part of the bodie that is nourished, there is left some part of profitable bloud, not needefull to the partes, ordeyned by nature for procreation, which ... is woonderfullie conveighed and carried to the genitories, where by their proper nature that which before was plaine bloude, is now transformed and changed into seede.* (240)

Semen, thus, is "a form of white blood," as Berkeley maintains in *Blood Will Tell* (37). During copulation, both men and women influence each other by mingling their sperma. In *The History of Biology* F. S. Bodenheimer elucidates Aristotle's conception: "Male and female particles mingle and they both exercise their influence, according to their relative strengths, transmitting characteristics of structure, of function, and of behaviour in the developing young" (55). It is not clear when Tamora begins to mingle her sperma with Aaron's semen. Yet one can surmise that their relations have developed since they met in the land of the Goths, on the grounds that Aaron was brought to Rome along with the Queen and that in his soliloquy Aaron in a sexual innuendo gloats both over Tamora's ascension to the Roman Empery and over his sexual relationship with her:

> *Then, Aaron, arm thy heart and fit thy thoughts*
> *To mount aloft with thy imperial mistress*
> *And mount her pitch whom thou in triumph long*
> *Hast prisoner held, fettered in amorous chains*
> *And faster bound to Aaron's charming eyes*
> *Than is Prometheus tied to Caucasus.* (II.i.12–17)

Therefore, through their continuing clandestine dalliance, Aaron's base white blood (semen) is mingled with Tamora's royal sperma, which leads to the degeneracy of her ethical standards. Francis Markham in *The Book of Honour* maintains that noble blood mingled with base blood produces "an imperfect generation," and that "many generations are needed to raise a gentleman of good quality" (47–48). The "imperfect generation" of Aaron and Tamora is the blackamoor child or, as the nurse describes it, "A joyless, dismal, black, and sorrowful issue! … the babe, as loathsome as a toad" (IV.ii.67).

Tamora's villeinization by Aaron the Moor parallels that of the evil Queen in *Cymbeline* by her first nameless—almost surely plebeian—husband and that of Queen Gertrude in *Hamlet* by the villein King Claudius. Considering the ages of the children in these plays—Demetrius and Chiron (of Tamora), Cloten (of the evil Queen), and Hamlet (of Gertrude)—one may easily conclude that these parents are in their late forties, if not fifties. Despite these women's status as widows, probably because they are still beautiful, they can still charm their new husbands (Saturninus, Cymbeline, and Claudius). Though comparatively old, these Queens are still beautiful and still speak in blank verse, both of which are signs of gentility. Berkeley and Karimipour argue that "Class-originated beauty is usual in Shakespeare's plays" (92). They regard Perdita's singular beauty as "an effect of her high blood" (91). Admittedly, all of the Shakespearean heroines are gentle of blood and remarkably beautiful: Silvia, Perdita, Juliet, Imogen, Rosalind, and Portia, to name only a few. And they speak in blank verse, another sign of their gentility. Berry observes that "Class identification is confirmed through language …. As a general rule, blank verse is the natural medium of gentry, as with nobility and royalty. It is the language of passion, dignity, and moral elevation, hence is equated with social elevation" (xv–xvi). But their gradual villeinization takes place if they have intercourse with their villeinous husbands. In his article "Claudius the Villein King of Denmark," Berkeley argues that "night after night she [Queen Gertrude] is becoming less consanguineous with him [the elder Hamlet, her first husband] … and becomes villeinized with Claudius' thick, cold, sluggish, ill-tasting, darkish blood (and therefore acquiring all his vicious and ugly qualities)" (9). In the same manner, Cloten's mother seems to have been much villeinized with her first nameless husband's base blood. Tamora is no exception. From the outset of the play, her moral depravity is evident in her role as a Machiavel (to the Andronici), as an enchantress or a witch (to Saturninus), as a mistress (to Aaron, an adulterer who makes Saturninus a cuckold), as an avenger (to the Romans in general and to the Andronici in particular), as a mother of a bastard (to a blackamoor), and as an encourager of rapists and murderers (to her sons). All of these evildoings are closely related to Aaron and are putative results of Tamora's villeinization by him. Her villainy is stopped only when Titus stops her life. She deserves no heavenly hope or vision of rebirth.

Tamora's corrupt morality has a bad effect on her new husband, Saturninus. When Bassianus and Lavinia enter again, now as husband and wife, Saturninus' pride is hurt and his desire for vengeance is renewed. His royal blood makes him angry, or choleric—a gentlemanly temper—at their marriage against his wish. Tamora pretends to plead with him to "pardon what is past"(I.i.432). The naive Saturninus shouts, "What, madam? Be dishonoured openly / And basely put it up without revenge?" (433–34). He is still so strongly conscious of his royal blood as to seek open and honorable revenge on the Andronici and Bassianus for taking Lavinia away from him. In an aside, however, Tamora teaches him a Machiavellian policy: "My lord, be ruled by me, be won at last; / Dissemble all your griefs and discontents .... I'll find a day to massacre them all" (443–44, 451). Like the degenerate King Richard III and the evil Queen of *Cymbeline*, Tamora is notorious for her Machiavellian dissembling. Her perverted character reflects her debased blood. Tamora exercises her vicious influence on the Emperor night after night, while she still keeps her secret rendezvous with Aaron. Bewitched by Tamora's beauty and ineffable charms (mostly coming from her witchcraft, like the evil Queen of *Cymbeline* or the witch-Queen in the fairy tale *Snow White*), Saturninus fails to recognize his own villeinization and the consequent corruption of his empire. Like Tamora, he deserves no vision of heavenly rebirth. His death at the hands of the ideal hero, Lucius, marks the end of corrupted Rome.

Since Saturninus' marriage to Tamora, the Empress and her paramour Aaron, collaborators in the Emperor's villeinization, have emerged as threats to the Roman Empire. Leslie Fiedler in *The Stranger in Shakespeare* argues that Shakespeare presents Jews, witches, blacks, and savages as strangers who threaten destruction rather than offer hope of salvation, literal or metaphorical (15). Thus Tamora as a Gothic witch and Aaron as a Moor are strangers who threaten destruction of the long-descended Andronici as well as of the Emperor and his brother—and, by implication, of the whole Roman Empire. These strangers are mostly base by birth and do evil deeds, which destroy some innocent victims, but Shakespeare never lets baseborn strangers enjoy his blessing at the close of his plays. With no exception, they are all punished by being humiliated or even killed by bloods. And the poet generally has the ideal hero or heroine restore peace and order to society, not solely in the comedies but also in the tragedies and the histories.

Admittedly, Shakespeare almost always dramatizes the bipolar contention between the two classes—armigerous and plebeian. In *Titus Andronicus* the Andronici epitomize the armigerous class, and Aaron the Moor exemplifies the evil plebeian. Titus takes pride in his honored family whose "monument five hundred years hath stood" (I.i.351)—the longest regularly descended family in the Shakespearean canon. Titus' glorious victory over the Goths heightens the fame and the pride of his family. But Aaron the Moor not merely threatens destruction of his family but also effects the villeinization of Tamora and her traitorous sons through Machiavellian policy and violence. In fact, these Goths and Saturninus join Aaron in competing with the gentle Andronici. Especially by highlighting the blackness of Aaron's skin and character, Shakespeare maximizes the visual and thematic effects. G. K. Hunter in "Othello and Colour Prejudice" states that the Elizabethans have the basic and ancient prejudice against the color black: they think of black as "the colour of sin and death" (182). In *Love's Labor's Lost,* the King of Navarre declares: "Black is the badge of hell, / The hue of dungeons and the school of night" (IV.iii.250–51). Aaron's blackness suggests not only his devilish character and his inferior blood; as a proverb says, "Three Moors to a Portuguese; three Portuguese to an Englishman."[5] Most Elizabethans seem to have a hierarchy of nations in mind when they distinguish between England and other countries such as the lands of Goths, Moors, and Turks, in accordance with the dictum of Francis Markham: "there are severall degrees in bloud" (46).

Not only Aaron's black appearance but also his base humor—"my cloudy melancholy" (II.iii.33)—suggests his base blood. Aaron's "cloudy melancholy" and Edmund's "villainous melancholy" (*King Lear* I.ii.138) exemplify Galenic melancholy as a disease of the non-gentles or of gentle illegitimates. As Berkeley delineates it, Galenic melancholy differs from the "good melancholy" of Bertram's father in *All's Well That Ends Well* I.ii.56, "the fashionable pseudo-Aristotelian melancholy" (*Blood Will Tell* 9)—a dominant humor in the gentle characters like Bertram's father, Olivia, Orsino, Viola, Romeo, Jaques, and Hamlet.

Unlike the gentles, plebeians like Aaron and Shylock and the degenerate gentles like Richard III, and illegitimate offspring such as Edmund, suffer from Galenic melancholy and, as John W. Draper holds, are "all in revolt against established order, and therefore are conspirators, usurpers, and villains" (63). During the Middle Ages, Galenic melancholy was linked to the planet Saturn, and phlegm was related to Venus. The astrological belief became widespread that the particular planet ascendant at the time of a person's birth influenced his or her temperament. Irving I. Edgar explains that Saturn was believed to cause excessive black bile, promoting melancholy, and Venus, associated with the phlegmatic humor, was "thought proper to women, children, and voluptuaries."[6] The influences of Saturn and the melancholic humor on Aaron engender his vengeful spirit, as he confides it to Tamora, a woman of Venus or phlegm:

> *Madam, though Venus govern your desires,*
> *Saturn is dominator over mine.*
> *What signifies my deadly-standing eye,*
> *My silence, and my cloudy melancholy, ...*
> *No, madam, these are no venereal signs.*
> *Vengeance is in my heart, death in my hand;*
> *Blood and revenge are hammering in my head.*
> (II.iii.32–9)

Aaron's vengeful spirit partly results from his Galenic melancholy—the humor of baseborns—because Saturnine men "will never forgive till they be revenged,"[7] and it partly results from his jealousy because his mistress became Saturninus' empress. The first two lines contrast Tamora's erotic desire and Aaron's melancholic and vengeful mood. The phlegmatic women under Venus, as C. Dariot in *Astrologicall Iudgement of the Starres* suggests, are "louers of delights."[8] "Louers of delights" or "voluptuaries" are like Tamora and the evil Queen of Cymbeline; they sharply contrast with true romantic lovers like Juliet, Portia, and Lavinia. The evil Queen and Tamora are not unlike the "wanton Maidens" and Malecasta, the lustful Lady of Castle Joyeous whose "hasty fire" and "fickle heart" are devoid of the sense of chastity in the second and the third books, respectively, of Edmund Spenser's

The Faerie Queene. Being melancholic and revengeful, Aaron restrains Tamora's carnal desires: "No, madam, these [melancholic traits] are no venereal signs." He knows that her sons also inherit "that codding spirit" from their mother (V.i.99). Melancholy and phlegm are base humors in contrast to the gentle humors, blood and choler; Aaron's melancholy signifies his lowly birth, and Tamora's phlegm implies her villeinization, as Aaron perceptively connects her to the planet Venus. (Even Saturninus belongs to the planet Saturn and hence to melancholy, because, as Eugene M. Waith suggests, the name reflects Shakespeare's interest in the astrological theory that saturnine men were "false, envious ... and malicious."[9] His name, then, seems to imply his villeinization as well as his inability to excercise justice as Emperor.) The Moor compares himself to the poisonous snake which is ready "To do some fatal execution" (II.iii.36). Thomas Walkington in his *Optick Glasse of Humors* connects Saturnine men with "dangerous Matchiavellisme" (129): Aaron also reveals his Machiavellian policy to vent his fury and desire for revenge.

The major target of Aaron's villainy is, of course, the Andronici, but virtually every Roman, including Bassianus, falls victim to it. Thus Marcus concludes that Aaron is the "Chief architect and plotter of these woes" (V.iii.122). Throughout the play the Moor has been spiteful and revengeful toward the Romans, who are gentle by birth, sanguine or white in color, and Roman in citizenship. Aaron rages when Demetrius and Chiron attempt to kill his blackamoor son, and he launches a harsh attack on their sanguine temperament and white skin: "What, what, ye sanguine, shallow-hearted boys! / Ye white-limed walls! Ye alehouse painted signs! / Coal black is better than another hue" (IV.ii.98–100). He goes on to condemn the white complexions of Chiron and Demetrius as being of "treacherous hue" because they "will betray with blushing / The close enacts and counsels of thy heart!" (118–19). Similarly, later as a captive of Lucius, Aaron takes pride in being black, referring to a proverb—"[To blush] like a black dog" (V.i.122). Aaron also has so strong an aversion to Romans that he has dug up dead men from their graves and on their white skins carved with a knife in Roman letters: "Let not your sorrow die, though I am dead" (135–39).

Aaron vents his strong antagonism against the Romans by encouraging Tamora's sons to ravish Lavinia, a representative gentlewoman of Rome. Demetrius and Chiron vie for Lavinia; they even draw rapiers and quarrel over her. The sibling rivalry between the Roman imperial brothers is now parodied by a kind of mock sibling rivalry between the lustful Gothic brothers. Saturninus' and Bassianus' arguments for their claim to the crown parallel the ludicrous and fallible reasoning of Demetrius and Chiron. Chiron, the younger brother, belittles the differences in age and emphasizes ability:

> *'Tis not the difference of a year or two*
> *Makes me less gracious or thee more fortunate;*
> *I am as able and as fit as thou*
> *To serve, and to deserve my mistress' grace.* (II.i.31–34)

And he demonstrates his resolution: "I care not, I, knew she and all the world. / I love Lavinia more than all the world" (71–72). On the other hand, Demetrius stresses that he is the elder brother: "Youngling, learn thou to make some meaner choice. / Lavinia is thine elder brother's hope" (73–74). He goes on to explain how he can win Lavinia in a quasi-syllogism:

> *She is a woman, therefore may be wooed;*
> *She is a woman, therefore may be won;*
> *She is Lavinia, therefore must be loved.* (82–84) [10]

While Tamora teaches Machiavellian policy to Saturninus, Aaron directs Demetrius and Chiron to use Machiavellian "policy and stratagem" in order to satisfy their lust for Lavinia by raping her in turn (II.i.104). This heinous scheme they pursue, the elder brother first and the younger brother next. Thus Tamora and her sons are under the Moor's evil influence, although of course in respectively different ways. Later captured by Lucius, Aaron confesses, "Indeed, I was their tutor to instruct them [in rape, mutilation, and murder]. / ... That bloody mind I think they learned of me" (V.i.98,101).

Because of his essential villainy and blackness, Aaron attracts many critics' attention. Bernard Spivak regards him as a descendant of the Vice figure of medieval allegorical drama (380). Muriel C. Bradbrook stresses his blackness as an "outward symbol of his diabolic nature" by comparing him both to the medieval Vice and to the "conscienceless Machiavel," thus making Aaron the precursor of Richard III and Iago (107). H. Bellyse Baildon sees Aaron as a crude anticipation of Iago, Regan, Goneril, and Richard III, each of whom is an embodiment of a "lost soul" (xliii). Fredson Thayer Bowers compares him to the Marlovian protagonist and villain Barabas (118). Irving Ribner considers him "the manipulator of the evil action, the specific author of Titus's misfortunes … a symbol of evil itself" (18).

But none of these critics examines Aaron as the main cause of Tamora's villeinization and as the leader of the devil's party. Therefore, they fail to perceive Aaron's influence on Tamora and their close relationship, and thus they miss many dramatic effects Shakespeare aims at. Bowers, for example, rightly considers Aaron the central villain from Act Two onward, but he faults Shakespeare for diverting attention from Tamora (the original instigator of the villains' revenge) and thus disturbing the "symmetry of the plot" (118). The shift of emphasis from Tamora to Aaron by no means destroys the structural unity of the play, but instead it intensifies the bipolar contrast and conflict between the two classes, gentle and base. Indeed, most of the base and evil deeds of Tamora and her brood—cuckoldry, rapes, mutilations, murders, and other atrocities—hinge on Aaron. Not only does he offer vicious counsels to them, but he also causes Tamora's gentle blood to be "cold, slow-flowing, sour-tasting, dark, and heavy with phlegm," all of which are characteristic of villein blood (Berkeley and Karimipour 89). A product of her villeinization is the "imperfect generation"—a blackamoor child, or "the base fruit of her burning lust," as Lucius puts it (V.i.43).

Aaron's only trait of behavior which might earn the reader's sympathy is his natural sense of protecting his blackamoor baby. Aaron is the only villain in Shakespeare indisputably to have a child, though it is black and ugly. Just as Titus' ability to beget many children (twenty-five sons and a daughter) is a strong testimony of his rich and abundant blood, Aaron's begetting a child indicates that his own blood is somewhat rich. When Lucius taunts the blackamoor baby as "the fruit of bastardy," Aaron shouts, "Touch not the boy. He is of royal blood" (V.i.49). Like Cloten in *Cymbeline*, who erroneously thinks of his blood as royal because his mother is the King's wife, Aaron falsely believes the baby's blood to be royal because his mother is Rome's Empress. Although the baby's blood is not royal—maybe slightly gentle—Aaron desperately tries to protect this child. His protectiveness toward his own child is noteworthy because it implies that his heart is not totally inhuman. This may reflect Tamora's positive influence on him; that is, some of Tamora's initially gentle blood is transmitted through her sperma to Aaron's base blood and somewhat ameliorates it, as "Male and female particles mingle and they both exercise their influence" (Bodenheimer 55). Tamora's earnest pleas for mercy on her son, Alarbus, in the opening scene of the play may deserve the audience's sympathy, and this, I think, is a sign of her initial gentility. And her order to kill the blackamoor shows her degeneracy. Tamora's degenerate blood does not allow her to repent of her evildoings; she is punished by being stabbed by Titus. Owing to his innate base blood, Aaron rejects any possible opportunities to repent of sins and delivers these last words: "If one good deed in all my life I did, / I do repent it from my very soul" (V.iii.189–90). He is punished with a most humiliating and painful death: being captured by a soldier of Lucius' army, he is condemned to death by being set breast-deep in earth, there to die by starvation. Aaron's unrepentant last words parallel those of the evil Queen of Cymbeline. The degenerate Queen also repents lost opportunities to effect her villainy: Cornelius reports that before death she grieves that "the evils she hatch'd were not effected" (*Cymbeline* V.v.60).

Aaron and Tamora thus represent the party of strangers whose blood is villeinish or villeinized, as opposed to the Andronici, the pure Roman bloods. The conflict between gentle and villein is visualized concretely in the literal opposition between Aaron's blackness of skin and morality—"Aaron will have his soul black like his face" (III.i.205)—and Lavinia's assailed whiteness of body and spirit. The differences between black and white, dark and light, barbarian and civilized, evil and innocent specify the bipolar opposition between the evil, baseborn Aaron and the innocent gentlewoman Lavinia. The floral imagery reinforces this binary opposition, especially when the innocent Lavinia is victimized by Aaron's villainy in the forest. According to Aaron's "policy and stratagem," Demetrius and Chiron ravish Lavinia and cut off her tongue and hands in the dark forest, which as a microcosm embodies the corrupting of the Roman Empire by the evil villeins: as Titus groans, "Rome is but a wilderness of tigers" (54).

The garden and the forest are Shakespeare's favorite metaphors for society, and an individual tree or flower often embodies a family or a single person. In his first soliloquy, Hamlet pictures the corrupt Danish court as the "unweeded garden" which is overgrown by weeds—corrupt courtiers led by his uncle Claudius, who usurped his brother's throne and committed incest (hence "incestuous sheets") with his sister-in-law. In *Titus Andronicus* the prosperous and peaceful Rome deteriorates into a wilderness full of animals of prey. The fair trees and flowers (innocent victims—Lavinia and Bassianus) lose their limbs and even their lives by the "ungentle hands," and the unfortunate prey with its brood (the gentle Titus and his children) awaits his doom by tigers (the devil's party led by Aaron). In Albert H. Tricomi's expression, the devil's party are "the panther and the tigress with her whelps overruning the forest," destroying "its initial pastoral identity" (100).

Copious floral images, which poetically characterize the gentle and the base metaphorically, reflect the poet's understanding of the prevailing Elizabethan sense of hierarchy. The Elizabethans used to categorize various kinds of trees, fruits, and flowers according to their literal and symbolic characteristics. Sir Walter Raleigh in his Preface to *History of the World* puts the cedar on the highest rank and the shrub on the lowest. Henry Peacham in his opening chapter of *The Compleat Gentleman* identifies the oak tree as "the forest's king," esteems the rose most of all the flowers, and admires "the pomeroy and queen-apple" among other fruits. Likewise, Shakespeare links the gentles to the highly valued objects corresponding to their human counterparts. In contrast, the shrubs and weeds belong to all seasons except winter, and they represent the base on many occasions. The garden flowers, roses and lilies, stands on the pinnacle of the hierarchy of flowers partly because of their remarkable fragrance and beauty on the literal level, and partly because they are symbolic of chastity, beauty, and purity—gentlewomen's traits. In other words, lilies and roses, among other garden flowers, symbolize beautiful and chaste gentlewomen both in terms of physiognomy and figure, and in the morality and ethics of the Elizabethan period. In *Titus Andronicus,* when Marcus first notices Lavinia's injuries in the forest, he asks his "gentle" niece which "ungentle hands" mutilated her limbs and tongue. He uses floral imagery:

> *Speak, gentle niece, what stern ungentle hands*
> *Hath lopped and hewed and made thy body bare*
> *Of her two branches, those sweet ornaments,*
> *Whose circling shadows kings have sought to sleep in ....*
> (II.iv.16–19)

Marcus associates his gentle niece with a tree and flowers which accord with her femininity and gentility. The tree's branches offered "circling shadows" to the gentle of blood—kings being the primates—and the intimacy between the tree and kings is evident. He describes how the gentle blood of the "deflowered" Lavinia flows down between her "rosed lips," and he also reminisces how her "lily hands" trembled like "aspen leaves" upon a lute—a musical instrument of the gentle—before the mutilation (44–45). Shakespeare's plays and poems abound in these kinds of floral images. In *The Rape of Lucrece,* before Tarquin steals into Lucrece's chamber and ravishes her, Lucrece sleeps with "Her lily hand her rosy cheek lies under" (line 386). Rosy cheeks always import life and virginity in Shakespeare's works. Friar Laurence explains how his distilling liquor will cause the roses in Juliet's lips and cheeks to fade away (*Romeo and Juliet* IV.i.99). Othello regrets that he "plucked the rose [Desdemona]" (*Othello* V.ii.13). Guiderius says, on seeing what he supposes to be Imogen's dead body, "O sweetest, fairest lily!" (*Cymbeline* IV.ii.203). Lilies, a symbol of purity and chastity, Shakespeare himself relates to gentry, while the base he figures as weeds (Sonnet 94).

Marcus describes Lavinia's gentle blood as rich, warm, red, and fast-flowing: "Alas, a crimson river of warm blood, / Like to a bubbling fountain stirred with wind" (II.iv.22–23). Just as King Duncan's abundant blood despite his old age horrifies Lady Macbeth—"Yet who would have thought the old man to have had so much blood in him?" (V.i.37–38)—the ample blood of Lavinia even in a state of triple amputation astounds Marcus:

And notwithstanding all this loss of blood,
As from a conduit with three issuing spouts,
Yet do thy cheeks look red as Titan's face
Blushing to be encountered with a cloud. (29–32)

Like such innocent victims as Ophelia, Cordelia, Desdemona, Duncan, Humphrey, Talbot, and Bassianus, Lavinia fails to survive evil forces—agents of social corruption and destruction. Her death, however, will be enshrined in her household monument and mourned by the new Emperor Lucius, her brother. It is a pagan form of rebirth. By contrast, as for Tamora, the "ravenous tiger," the Emperor orders, "No funeral rite, nor man in mourning weed, / No mournful bell shall ring her burial; / But throw her forth to beasts and birds of prey" (V.iii.196–8). Tamora suffers a pagan form of damnation.

Titus Andronicus shares the pivotal role with Aaron in the conflict between the gentle Romans and the Gothic and Moorish strangers. Opposed to the devil's party inspired by Aaron, Titus plays a central role in the Andronici and the Roman Empire as the head of the Andronici and the general of the Roman army. Criticism has been split about Titus' character. Baildon thinks of Titus, in the light of his tragic faults, as a rough pattern for the characters of Lear, Coriolanus, and Hamlet.[11] Ribner also regards Titus as a forerunner of Coriolanus, but in terms of his virtues; Ribner maintains that, like Coriolanus, "Titus embodies all the ancient Roman virtues." When exaggerated, Titus' virtues become faults, and his destruction is guaranteed by his failure to alter his behavior and accept salvation: "He is a great and initially Virtuous man, the first of Shakespeare's heroic figures whose very virtues are the source of their sins." In the life journey of his hero, Ribner surmises, "Shakespeare explores in imaginative terms the universal way of damnation, for Titus becomes a prototype of erring humanity" (17). E. M. W. Tillyard approvingly examines the Janus-faced character of Titus in his *Shakespeare's History Plays*: in his madness, Titus comes close to Kyd's Hieronymo, but in his sanity he is "an elderly Talbot"—a brave warrior and an unswerving servant of his royal master (139). By contrast, H. B. Charlton depreciates the role of Titus, regarding him as the "nominal hero" while considering the villains, Aaron and Tamora, to be the real protagonists (21): indeed, they light up several notable facets of Titus' character. The diversity of criticism reflects Titus' dual character, a mixture of gentility and degeneracy. But the critics ignore his pivotal role as the blood-oriented revenger who seeks vengeance on the devil's party, which consists of the base and villeinized strangers inspired by Aaron.

The degeneracy of Titus darkens the future of the whole Roman society. He appears in the opening scene as the quintessential virtuous gentleman. When Titus returns home victorious from a long war against the Goths, the Romans unanimously select him as their new emperor because, as Marcus sums up, "A nobler man, a braver warrior [than Titus], / Lives not this day within the city walls" (I.i.25–26). A captain also eulogizes him: "The good Andronicus, / Patron of virtue, Rome's best champion, / Successful in the battles that he fights, / With honor and with fortune is returned" (64–67). He lost twenty-one out of "five-and-twenty valiant sons," and his family tomb becomes the "Sweet cell of virtue and nobility" (93). His victory over the Goths secures Rome's prosperity, and the bond of his family mirrors the order of the Empire. However, the family union of the Andronici and the order of Rome collapse, both because of Titus' self-subverted retirement in favor of the Saturnine man, and because of his adamant espousal of primogeniture and his ruthless exercise of patriarchal authority. The familial discord surfaces first when Titus' sons back Bassianus' claim to Lavinia's hand in spite of the new Emperor's proclamation that he will marry the maid. Titus becomes furious, not only because the sons rebel against his patriarchal authority, but also because he prefers the firstborn son to the second son as his son-in-law. In Titus' mind, it is intimated, to keep the traditions of primogeniture and patriarchal authority—that is, to support the blood-based hierarchy—is to guarantee the order and peace of his family as well as of the state. Titus also regards breaking these traditions as dishonorable to the noble blood of his family. So Titus becomes angry at his daughter's betrothal to Bassianus without his approval, partly because Bassianus is not the firstborn prince. And he hates the idea of burying Mutius, who brandished the sword before his father in an attempt to help Lavinia marry Bassianus, in the "monument [where] five hundred years hath stood" (I.i.351). His view seems sound but not infallible. In the Elizabethan era, for a daughter to give a pledge to a man without her father's approval and for a son to lift up a sword against his father are equal to treason against the state and to Satanic rebellion against God. The Mosaic law, for example, prescribes death for striking a parent. These three types of rebellion equally represent the violation of the hierarchical order in a family, in a state, and in Heaven. In this sense, Titus acts like an Elizabethan father and courtier.

But Shakespeare does not servilely embrace these Elizabethan standards. Shakespeare's plays present many romantic lovers who betroth themselves without paternal approval—Bassianus and Lavinia, Othello and Desdemona, Romeo and Juliet, Lorenzo and Jessica, Lysander and Hermia, Florizel and Perdita, and Posthumus and Imogen. They withstand the fathers' threats, confinements, and banishments; they even run away from home for love. Denton J. Snider, a nineteenth-century critic, observes:

Shakespeare everywhere justifies the right of the daughter's choice when it is the sole issue, because it "belongs to the woman to say who shall be her husband, for she, and not her father, has to form with him the unity of emotion which lies at the basis of the Family .... He [Shakespeare] always mediates such a conflict by the triumph of the daughter .(88)

With regard to betrothal, Shakespeare seems to subordinate paternal authority to mutual love within the same class: these lovers are all gentle of blood. In the Shakespearean canon there is no cross-class marriage. If not of a cross-class, it appears, the betrothal of the lovers deserves authorial blessing as well as the audience's approval, with or without paternal endorsement. In brief, in Shakespeare's plays, rebellion against paternal authority can be excused, only if it is for the sake of mutual love within the same class—an integral harmony or order in a family. A rigid exercise of paternal authority often thwarts a prospective love (Polonius' influence in Ophelia's love for Hamlet and the incestuous Antiochus' control of his daughter's possible match with Pericles) and is often partly responsible for the deaths of the lovers (for example, Romeo and Juliet).

As the plot develops, Titus' rigid insistence on patriarchal control in the matter of his daughter's marriage turns out to be a delusion or a fancy. Attached to this illusory perspective, Titus disapprovingly calls Mutius the "villain boy" and stabs him as he tries to hinder Titus' own pursuit of Lavinia and Bassianus. To be sure, the word *villain* is a common curse in Elizabethan and modern English, but in Shakespeare the word is interchangeable with *villein.*[12] Titus thinks of his son's rebellion against him as a villeinous/villainous act. In other words, he condemns him as a villeinized bastard: he no longer thinks of him as his son. So he repudiates the pleas of his other sons to bury Mutius in the centuries-old family tomb:

Traitor, away! He rests not in this tomb.
This monument five hundred years hath stood,
Which I have sumptuously re-edified.
Here none but soldiers and Rome's servitors
Repose in fame, none basely slain in brawls.
Bury him where you can, he comes not here. (I.i.350–55)

Only after his sons and his brother Marcus plead for mercy does Titus reluctantly allow the burial. Titus' disowning Mutius echoes Tamburlaine's cold-blooded murder of his own cowardly son

Calyphas, who plays cards with his soldier Perdicas instead of joining his brothers. Tamburlaine refers to Calyphas as "this coward villain, not my son, / But traitor to my name and majesty" (*2 Tamburlaine* IV.i.91–92). He calls Calyphas and Perdicas "ye base, unworthy soldiers" (102), and stabs his own son, because in him was "neither courage, strength or wit, / But folly, sloth, and damned idleness" (128). Despite the Marlovian egalitarianism—Tamburlaine himself was a villein/villain because he was a Scythian shepherd—Tamburlaine as Emperor is suddenly conscious of his class.

In these similarly bloody scenes, both Tamburlaine and Titus clearly become the protagonists with whom the audience cannot identify or for whom the audience cannot feel pity and fear. The audience's detachment from them becomes more pronounced as they are associated with lunacy and damnation. Orcanes calls Tamburlaine's murder "thy barbarous damned tyranny" (IV.i.152); Quintus thinks his father, Titus, "is not with himself" (I.i.369). Titus' murder of Lavinia also is analogous to Olympia's honorable murder of her son and her own suicide to prevent the Scythians from tyrannizing over them (*2 Tamburlaine* III.iv.18–33). Just as Tamburlaine's family and his Empire decline rapidly after the murder of Calyphas, Titus' family and Rome decay promptly after the murder of Mutius. Titus' wasting of his two children epitomizes his failure as a father figure. His lack of capability as a patriarch is proven again by the failure of his role as the surrogate father to Saturninus, who calls him "noble Titus, father of my life!" (I.i.254), whom Bassianus also refers to as "a father and a friend to thee [Saturninus] and Rome" (424). Now Saturninus rejects the Andronici: "No, Titus, no. The Emperor needs her [Lavinia] not, / Nor her, nor thee, nor any of thy stock" (300–1). Titus' failure in playing a father figure hints at the collapse of Roman hierarchical order; the discord of the Andronici as a representative noble Roman family foreshadows the disorder of the Roman Empire. Saturninus' immediate rejection of Lavinia after a dispute and his "sudden choice" of the degenerate Tamora anticipate the speedy downfall of the Empire. Titus' hasty retirement and speedy nomination of Saturninus as the Emperor, along with Saturninus' "sudden choice" of Tamora, lead to the quick corruption of Rome. Rape (Lavinia), mutilation (Lavinia and Titus), murder (Bassianus and Titus' two sons), and banishment (Lucius) occur in fast succession in Rome—"a wilderness of tigers."

In this wilderness, all the animals are so revengeful and spiteful that they prey on one another, and even the fiendish tiger Aaron moralizes, "Vengeance rot you all!" (V.i.58). Not until these ensuing tragic incidents does Titus discover the reality of human animals of prey. The discovery torments him so deeply that he falls into a state somewhat similar to Lear's insanity and Hamlet's feigned madness. But the difference between Titus and Lear lies in the former's vigorous attitude toward revenge: he decides to return "all these mischiefs" and "swears unto my soul to right your wrongs" (III.i.273, 278). Although Tamora and her sons suppose him to be insane, Titus, like Hamlet concerning Claudius, knows all about them now. As the banished Lucius wins the hearts of the Goths and directs them to advance against Rome, events finally conspire to prompt Titus' revenge in the weird scene in which Tamora and her sons, disguised as Revenge, Rape, and Murder, come to Titus, who, they believe, has become demented. Now, however, Titus knows their reality—their dissembling nature, as he recapitulates: "I know thee well enough" (V.ii.21); "I know thee well" (25); "I know them all" (142). Martha Tuck Rozett observes that, as in *Hamlet*, there exists in *Titus Andronicus* an "ironic and accidental quality" to the circumstances leading to revenge: Tamora, though pretending to be a temptation figure, does in fact become one, and she, through providential intervention, falls into becoming the agent of her own destruction, and Titus turns out to be the instrument of revenging gods (198). The tragic hero is now completely transformed into a bloody revenger—he cuts the throats of Demetrius and Chiron, bakes their flesh in a pie, and serves it to their own mother, Tamora.

Should Titus' bloody revenge be condemned or tolerated, or, as Bacon terms it, is it "a kind of wild justice" or "the most tolerable sort" of revenge? This question is not easily answered because the question about the validity of revenge is very delicate. For example, in *His Practice in Two Bookes* Vincentio Saviolo approvingly asserts that "the revenge ought to be done honorably"; on the other hand, he holds that one has to avoid vengeance even if one's own wife has been ravished by an offender because "God, who (as S. Paul saith) will judge the Adulterer, will by means thereof give most severe judgement."[13] If Titus avenges the adultery forced upon Lavinia and the murder of Bassianus simply to assuage his own wrath, it would be wrong because mere revenge for its own sake is alien to the honorable code of the gentry. Max H. James argues that Titus "not only can but should seek to eliminate this evil in the name of the greater good of society" (37). To be sure, like other gentles in the Shakespearean canon, Titus demonstrates his sense of honor and courage by seeking revenge on those who ruined his family. His revenge is a product of his belief that failure to withstand the injustice and wrongs forced on his family is dishonorable to his gentle blood, and that such failure will lead to chaos and destruction in the family and the state. Titus' high blood makes him fight against evil and restore good. Especially in his role of noble avenger Titus and his tragic choices and actions deserve the audience's sympathy, but only after the deflowering and mutilation of Lavinia, his own mutilation, and the deaths of his two sons as the victims of Aaron's Machiavellian "policy and stratagem."

Yet the question still remains moot: Is Titus' bloody revenge on Tamora and her brood tolerable, either because "there is no law to remedy" for Titus, or because he actually eliminates the agents of evil "for the sake of the whole Roman society"? Christian teachings, Renaissance concepts of revenge, and classical ideas of honor and virtue cannot either condemn or condone Titus explicitly. Some may condemn him; some may justify him. The question, however, may be answered more clearly in relation to Titus' blood-quality. Degeneracy means not only corrupt morality but also diminution of the high blood in quantity as well as in quality. Titus' sorrows, tears, and worries consume the quantity of his noble blood. When the judges and senators take his two sons away to the place of execution, Titus weeps and says to them, "in the dust I write / My heart's deep languor and my soul's sad tears. / Let my tears stanch the earth's dry appetite" (III.i.12–14). As they do not hear him but pass by him, he laments, "In winter with warm tears I'll melt the snow / And keep eternal springtime on thy [earth's] face, / So thou refuse to drink my dear son's blood" (20–22). Jacques Ferrand in *Erotomania or a Treatise of Love* cites the opinion of Empedocles that "when any one was surcharged with any strong passion of the Mind, the Blood was troubled, and from thence followed Teares, in like manner as whey comes from Milke" (129). Hence, Romeo says to Juliet as he leaves her chamber, "Dry sorrow drinks our blood" (*Romeo and Juliet* III.v.59). In his edition of Shakespeare's works, Bevington explicates this passage: "The heat of the body in sorrow and despair was thought to descend into the bowels and dry up the blood." The Queen in *2 Henry VI* speaks of "blood-consuming sighs" and "blood-drinking sighs" (III.ii.61, 63). Titus says to his weeping grandson, "tears will quickly melt thy life away" (III.ii.51). The loss of the quantity of blood leads to degeneracy: the Queen of Henry VI soliloquizes, "Oft have I heard that grief softens the mind / And makes it fearful and degenerate" (IV.iv.1–2). Titus' tears are so abundant—as his blood is abundant (he has abundant blood enough to have begotten twenty-five sons)—that Marcus' napkin "cannot drink a tear of mine [Titus']" (III.i.140). The consuming of blood by weeping and lamenting results in the diminution of his blood. In order to avoid such diminution, the Queen of *2 Henry VI* says to herself, "Think therefore on revenge and cease to weep" (IV.iv.3).

When the revengeful Titus comes to know the agents of evil, he weeps no more. Instead, he declares, "I have not another tear to shed. / Beside, this sorrow is an enemy" (III.i.266), and he swears upon his soul to right their wrongs. But it seems that by "revenge" the Queen means *retribution,* which "suggests just or deserved punishment," whereas Titus means personal *retaliation.* Shifting from weeping to thinking of revenge signifies Titus' deeper degeneration—from physical to moral—from being a victim to being a revenger. Innocent victims like Talbot, Duncan, Ophelia, Lavinia, and Desdemona still keep their high blood; no one denies their heavenly vision of rebirth. But personal revengers who seek retaliation for injuries and wrongs deserve no heavenly vision. Marcus worries about the nature of Titus' revenge. He embraces what a sixteenth-century Christian, rather than a pagan Roman, might have embraced with regard to revenge. When Lavinia writes down the names of rapists, he wants to see "What God will have discovered for revenge" (IV.i.76). And when he sees the transformation of Titus from a weeper (victim) to a revenger, Marcus prays, "Revenge the heavens for old Andronicus!" (131).

Unfortunately, Marcus cannot prevent his brother's further degeneration because Titus never listens to his choric voice. The audience may sympathize, if not identify, with Titus for stabbing his daughter to avoid her individual as well as her familial shame, but he deserves damnation for killing his son to uphold paternal authority and for his private and brutal retaliation against Tamora and her sons. His brutality is a product of his degeneracy. The causes of Titus' degeneracy are not so clear as those of Tamora and Saturninus, but, as mentioned earlier, the main cause seems to be the consuming of his gentlemanly humor and temperament due to his "bitter tears" and blood-consuming groans and sorrows. The excessive loss of humor and gentle temperament leads him to lose his temper. In other words, he loses balance in body as well as in mind. His monomaniacal involvement in brutal violence reflects his unbalanced mind and his degeneracy. And as a microcosm his transformation from innocence and gentility to corruption and degeneracy mirrors the degeneration of the Roman Empire. Thus Aaron's only reliable dictum—"Vengeance rot you all!" (V.i.58)—rings again.

But with a glimmer of hope *Titus Andronicus* closes, as a phoenix rising from its ashes. Lucius, ideal hero of the play, ascends as the new Roman Emperor chosen by the "common voice," as his father was at the beginning of the play. Titus would have been too tainted to rule the new Empire; his blood has to be purged and renewed in the frame of his son. A Gothic soldier does not distinguish Titus from Lucius, but thinks of them as one—"the great Andronicus." The Goth glorifies the family and, quite surprisingly, curses his former Queen, Tamora:

> *Brave slip, sprung from the great Andronicus,*
> *Whose name was once our terror, now our comfort,*
> *Whose high exploits and honorable deeds*
> *Ingrateful Rome requites with foul contempt,*
> *Be bold in us. We'll follow where thou lead'st, ...*
> *And be avenged on cursed Tamora.* (V.i.9–13, 16)

Earlier, when Saturninus exiled him without a cause, he promised Lavinia, "Now will I to the Goths and raise a power / To be revenged on Rome and Saturnine" (III.i.299–300). Now the ideal hero, Lucius leads the Goths to wreak revenge on Aaron ("the incarnate devil"), Saturninus (the villeinized Emperor), and the corrupt Empire ("Ingrateful Rome"). He captures Aaron and his blackamoor son, and he kills Saturninus when the degenerate Emperor stabs Titus, saying, "Can the son's eye behold his father bleed?" (V.iii.65). Not only does he avenge his father's death, but he also plays the role of magistrate: "There's meed for meed, death for a deadly deed!" (66)—which echoes the justification of revenge in Numbers 35.19. Now being chosen the new emperor, Lucius as the incarnate state punishes evildoers and amends the wronged: he orders his attendants to "Set him [Aaron] breast-deep in earth and famish him" (179), to bury the dead Emperor in his father's grave, and to enshrine his sister and father in his family's monument. As for the "ravenous tiger," Tamora, he orders, "No funeral rite, nor man in mourning weed, / No mournful bell shall ring her burial; / But throw her forth to beasts and birds of prey" (196–98). Thus peace and order are restored not only to the Andronici but to Rome. Being the weak and degenerate ruler, Saturninus foreshadows Richard II, who banishes Bolingbroke for an obscure reason but is later dethroned by him. As Lucius pauses over his father's body he describes Titus to his young son as a loving, paternal figure who danced his grandson on his knee and told him many a pretty tale. This is a loving and tender man whom the audience never really knew, the image mirroring the initial gentility of Titus before his degeneracy. This intimation seems to revive in the relationship between Lucius and Young Lucius. Now as the new patriarch of the Andronici and the new Emperor of Rome, Lucius epitomizes the Shakespearean gentleman. No one can object to the verdict that Lucius' revenges are tolerable and even justified, like Hamlet's murder of Claudius, the incestuous usurper.

The conflicts between blood-oriented revengers, therefore, structure Shakespeare's first tragedy, *Titus Andronicus.* These conflicts precipitate crude violence—ravishment, mutilation, carnage, and cannibalism—and end with the ascension of the ideal hero (Lucius) at the cost of degenerate gentles (Saturninus and Tamora) and the evil baseborn (Aaron). Poetic justice is thus maintained in accordance with Shakespeare's propensity to blood-consciousness. Admittedly, the characterization, structure, and themes of *Titus Andronicus,* like those of other plays by Shakespeare, are regulated by "an invisible but firm line between gentility and plebeianism," as Berkeley cogently argues in *Blood Will Tell* (95). Overlooking this thematic and structural unity of *Titus Andronicus,* many critics have condemned the play as "a heap of Rubbish" and even questioned its authorship.[14] Primarily because of the crude violence of the play, a number of critics have hesitated to admit that the "gentle" Shakespeare[15] could have invented this "jumble of horrors."[16] As compared with the mature tragedies, of course, *Titus Andronicus* lacks the depth and the subtlety of motivation underlying the major characters' actions. But in this play Shakespeare exhibits the ability to create dramatic effects by means of bipolar contrasts and oppositions between blood-based revengers, which shape the structure of the play. A close examination of major characters' blood-based revenges thus sheds light on elements of thematic and structural unity in the play that otherwise may be overlooked or ignored.

Most of the blood-oriented issues—such as conflicts between the gentle and the base, and conflicts emerging in matters of patriarchal authority, sibling rivalry, primogeniture, physiological and astrological references—seem to be solely Shakespeare's, for these issues were alien to the several sources of the play. Ovid's *Metamorphoses* provides the source for the rape, mutilation, and serving of human flesh in a meal: Tereus rapes his sister-in-law Philomel and deprives her of her tongue; Procne avenges his sister Philomel by serving Itys, her own son by Tereus, to Tereus in a meal. Seneca's *Thyestes* also gives Shakespeare a prototypical structure of sibling rivalry and revenge by the serving of human flesh in a banquet: Thyestes seduces his brother Atreus' wife and usurps his crown, which prompts Atreus to take revenge on his brother by murdering his two sons and serving their flesh to their parent in a banquet. These classical sources provided Shakespeare with ideas for bloody scenes, but they are reticent about blood consciousness. In many ways, they are not immediate sources for *Titus Andronicus.* Major critics have regarded the anonymous prose chapbook *The History of Titus Andronicus* (published in 1764) as closest to the lost main source of Shakespeare's play. Yet this prose chapbook includes no sibling rivalry between the brothers; the unnamed Emperor is old and has a nameless prince who falls in love with Titus' daughter Lavinia. Titus Andronicus, the captain-general of the Roman army, kills the King of the Goths and captures the Queen, Attava. Her sons still continue the struggle against Rome. The Emperor wants to marry Attava, not because he is enchanted by her beauty, but because he wants to make peace with the Queen's sons, Alaricus and Abonus. The nameless Moor has clandestine dalliance with Attava but does not play the pivotal role in the chapbook; hence, the prose work lacks the dramatic conflicts between the gentle and the base. The wicked Queen Attava lets her sons and the Moor kill the Prince in an attempt to make "her two sons emperors jointly." The prose work has nothing to do with blood-oriented revenges: Titus never mentions his pride in a five-hundred-year-long family legacy or his rigid paternal authority or his preoccupation with primogeniture. He appears a simple avenger of his daughter's rape by the Queen's evil sons. From such a simple plot, Shakespeare creates many dramatic effects by adding blood-oriented issues and blood-oriented revengers.

These blood-conscious characters and issues are the key not only to the thematic and structural unity of *Titus Andronicus* but also to Shakespeare's characterizations in later plays. Indeed, major characters in *Titus Andronicus* foreshadow many later Shakespearean characters: Aaron is a prototype of Iago and Edmund; Tamora, of Cymbeline's evil Queen; Lucius, of Hamlet; Titus, of Lear; and Saturninus, of Richard II. This play should therefore be read and studied more closely and approvingly, in relation to the poet's keen interest in the significance of blood.

Notes

[1] All the Biblical quotations in this article are from *The Geneva Bible: A Facsimile of the 1560 Edition* (Madison: U of Wisconsin P, 1969).

[2] All Shakespearean quotations are from David Bevington's 1992 edition.

[3] See Figure 15 of Irving I. Edgar's *Shakespeare, Medicine, and Psychiatry* and the front book jacket of Aubrey C. Kail's *The Medical Mind of Shakespeare.*

[4] Berkeley and Karimipour 89.

[5] M. P. Tilley, *A Dictionary of Proverbs* (Ann Arbor, 1950) M. 1132.

[6] Jupiter is associated with blood, the sanguine humor, which was considered appropriate to princes, accepted lovers, and the jovial and fortunate; the sun is related to yellow bile or choler, appropriate to rulers and self-willed women, and, in conjunction with Mars, to soldiers, roisterers, and drunkards. The moon is associated with mental illness, as the word *lunacy* implies. For detailed information see Edgar 215. See also the opening chapter of C. A. Mercier's *Astrology in Medicine* (London: Macmillan & Co., 1914); John W. Draper's "Humors" 259; and F. David Hoeniger 109.

[7] *Kalendar of Shepherds: Kalendar and Compost of Shepherds* (1518), ed. G. C. Heseltine (1930): 141–42. In *Examen des Ingenios* (1590), J. Huarte also asserts that some melancholic men are "blasphemers, wily, double, friends of ill-doing; and desirous of revenge" (147). See also Lawrence Babb, *The Elizabethan Malady* (East Lansing: Michigan State College P, 1951) 57–58.

[8] C. Dariot, *Astrologicall Iudgement of the Starres,* trans. F. Wither (London, 1583). Quoted in John W. Draper's *The Humors & Shakespeare's Characters* 37.

9 Eugene M. Waith quotes the characteristics of a Saturnine man—"false, envious ... and malicious"—from *Kalendar of Shepherds* in his edition of *Titus Andronicus* 83.

10 Demetrius' quasi-syllogism reminds one of Richard of Gloucester's similar reasoning when he decides to seduce Lady Ann, widow of Edward, in *Richard III*:

> *Was ever woman in this humor wooed?*
> *Was ever woman in this humor won?*
> *I'll have her, but I will not keep her long. (I.ii.230–32)*

11 Baildon states that both Lear and Titus fancy that they have a true and disinterested love for their children, from which the whole tragic situation arises; that both Coriolanus and Titus have the same military and warlike qualities; and that Titus has points of resemblance to Hamlet in regard to feigned madness (xxxiii–xxxv). Baildon's observations are sound but it is not clear whether Titus does feign madness, although Tamora and her sons obviously believe his madness to be real.

12 Bevington in a footnote suggests that Marcus' reference to Aaron as "the Empress' villain" (IV.iii.73) signifies "both servant and villain in the modern sense" (967). Berkeley also capitalizes on the interchangeability of the words *villein* and *villain* in his essay "Claudius the Villein King of Denmark" 9–21. For the origin and the development of the word *villain*, see Wilfred Funk 110–11.

13 Vincentio Saviolo, *His Practice in Two Bookes: The First Intreating of the Use of the Rapier and Dagger; The Second, of Honor and Honorable Quarrels* (London, 1595) Y4v and Zr. Quoted in Max H. James, "*Our House is Hell*": *Shakespeare's Troubled Families* (New York: Greenwood Press, 1989) 37.

14 Edward Ravenscroft, whose adaptation of *Titus Andronicus* was first performed in 1678, is the first critic to doubt its authorship. In the preface to his adaptation, he contends that he was once told by "some anciently conversant with the Stage" that it was written by an unknown author, and Shakespeare gave only "some Master-touches to one or two of the Principal Parts or Characters." Ravenscroft believes this because the play is "the most incorrect and indigested piece" in the Shakespearean canon, and its structure seems rather "a heap of Rubbish." Edward Ravenscroft, "Preface" to *Titus Andronicus*, rpt. Brian Vickers, ed., *Shakespeare: The Critical Heritage*, 4 Vols. (London and Boston: Routledge & Kegan Paul, 1974) 1:319.

15 Ben Jonson referred to Shakespeare as "My gentle Shakespeare" in "To the memory of my beloved, The Author Mr. William Shakespeare," which Jonson prefixed to the First Folio of 1623.

16 See Gareth Lloyd Evans, *The Upstart Crow: An Introduction to Shakespeare's Plays* (London: J. M. Dent & Sons Ltd., 1982) 39, and Oscar J. Campbell, ed., *The Reader's Encyclopedia of Shakespeare* (New York: Thomas Y. Crowell Co., 1966) 880. For more detailed discussions about the controversy over the authorship of *Titus Andronicus,* see pages 11–20 of Eugene M. Waith's introduction to his edition of *Titus Andronicus.*

Works Cited

Anderson, Linda. *A Kind of Wild Justice: Revenge in Shakespeare's Comedies.* Newark: U of Delaware P, 1987.

Aristotle. *Generation of Animals.* Trans. A. Platt. Vol. 5 of *The Works of Aristotle Translated into English.* Ed. W. D. Ross. Oxford: Clarendon P, 1908–1952.

Bacon, Francis. "Of Revenge." *Essays, Civil and Moral.* Ed. Charles E. Eliot. Vol. 3 of The Harvard Classics. New York: The Collier Press, 1909.

Baildon, H. Bellyse. "Introduction" to *The Works of Shakespeare: The Lamentable Tragedy of Titus Andronicus.* London: Methuen and Co., 1904.

Berkeley, David Shelley. *Blood Will Tell in Shakespeare's Plays.* Lubbock: Texas Tech UP, 1984.

—. "Claudius the Villein King of Denmark." *Hamlet Studies* 11 (1989): 9–21.

—, and Zahra Karimipour. "Blood-Consciousness as a Theme in *The Winter's Tale.*" *Explorations in Renaissance Culture* 11 (1985): 89–98.

Berry, Ralph. *Shakespeare and Social Class.* Atlantic Highlands: Humanities Press International, Inc., 1988.

Bevington, David, ed. *The Complete Works of Shakespeare.* 4th ed. New York: HarperCollins Publishers, Inc., 1992.

Bodenheimer, F. S. *The History of Biology.* London: Dawson and Sons, 1958.

Bowers, Fredson Thayer. *Elizabethan Revenge Tragedy, 1587–1642.* 1949.

Bradbrook, Muriel C. *Shakespeare and Elizabethan Poetry: A Study of His Earlier Work in Relation to the Poetry of the Time.* London: Chatto and Windus, 1951.

Broude, Ronald. "Revenge and Revenge Tragedy in Renaissance England." *Renaissance Quarterly* 28 (Spring 1975): 38–58.

Cogan, Thomas. *The Haven of Health.* London, 1584.

Draper, John W. "The Humors." *Journal of American Medical Association* 188 (1964):259.

___. *The Humors & Shakespeare's Characters.* Durham: Duke UP, 1954

Edgar, Irving I. *Shakespeare, Medicine, and Psychiatry.* New York: Philosophical Library, 1970.

Ferrand, Jacques. *Erotomania or a Treatise of Love*. London, 1640.

Funk, Wilfred. *Word Origins: An Exploration and History of Words and Language*. New York: Wings Books, 1992.

Galen, Claudius. *On the Natural Faculties*. Trans. Arthur J. Brock. Cambridge: Harvard UP, 1963.

Hoeniger, F. David. *Medicine and Shakespeare in the English Renaissance*. Newark: U of Delaware P, 1992.

Huarte, J. *Examen des Ingenios*. Trans. R. C[arew]. London, 1590.

Hunter, G. K. "Othello and Colour Prejudice." *Interpretations of Shakespeare*. Ed. Kenneth Muir. Oxford: Clarendon, 1985.

Kail, Aubrey C. *The Medical Mind of Shakespeare*. Sydney: Williams & Wilkins, 1986.

Lamb, Charles. "On the Tragedies of Shakespeare Considered with Reference to Their Fitness for Stage Representation." *The Works in Prose and Verse of Charles and Mary Lamb*. Vol. 1. Ed. Thomas Hutchinson. Oxford: Oxford UP, 1908. 124–42.

Markham, Francis. *The Booke of Honour: or Five Decades of Epistles of Honour*. London, 1625.

Peacham, Henry. *The Compleat Gentleman*. London, 1622. Rpt. *The Complete Gentleman, The Truth of Our Times, and The Art of Living in London*. Ed. Virgil B. Heltzel. Ithaca: Cornell UP, 1962.

Ravenscroft, Edward. Preface to *Titus Andronicus*. Rpt. Brian Vickers, ed. *Shakespeare: The Critical Heritage*. Vol 1. London and Boston: Routledge & Kegan Paul, 1974.

Ribner, Irving. *The English History Play in the Age of Shakespeare*. Princeton: Princeton UP, 1957.

Rozett, Martha Tuck. *The Doctrine of Election and the Emergence of Elizabethan Tragedy*. Princeton: Princeton UP, 1984.

Shakespeare, William. *The Complete Works of Shakespeare*. Ed. David Bevington. 4th ed. New York: HarperCollins Publishers, Inc., 1992.

Spenser, Edmund. *The Faerie Queene*. Ed. A. C. Hamilton. London and New York: Longman, 1987.

Spivak, Bernard. *Shakespeare and the Allegory of Evil Villains*. New York: Columbia UP, 1958.

Thayer, C. G. *Shakespearean Politics: Government and Misgovernment in the Great Histories*. Athens: Ohio UP, 1983.

Tillyard, E. M. W. *Shakespeare's History Plays*. New York: Barnes & Noble, Inc., 1964.

Van Doren, Carl. *The Oxford Dictionary of Quotations.* London: Oxford UP, 1941.

Vickers, Brian. *Shakespeare: The Critical Heritage.* London and Boston: Routledge & Kegan Paul, 1974. 4 vols.

Walkington, Thomas. *The Optick Glasse of Humors.* London, 1607.

Waith, Eugene M., ed. *Titus Andronicus.* Oxford: Clarendon Press, 1984.

Young, David. *The Action to the Word: Structure and Style in Shakespearean Tragedy.* New Haven and London: Yale UP, 1990.

Secondary Wisdom: The Role of Women as Mentors in Shakespeare's Plays

by John W. Crawford

The role of women in Shakespeare's plays has been examined in some detail since Anna Jameson's *Shakespeare's Heroines: Characteristics of Women* and Mary Cowden Clark's *The Girlhood of Shakespeare's Heroines*, both nineteenth-century tomes. That interest continued in the twentieth century with Agnes Mure MacKenzie's 1924 book *The Women in Shakespeare's Plays* and in Juliet Dusinberre's 1975 book *Shakespeare and the Nature of Women.* The latest entries include Angela Pitt's *Shakespeare's Women* and Irene G. Dash's *Wooing, Wedding, and Power: Women in Shakespeare's Plays.* As revealing as these studies are about the role of women in Shakespeare's plays, none treats the women as mentors for men. Yet a careful look at Shakespeare's drama demonstrates his recognition of Renaissance females as intelligent and capable of leadership.

Shakespeare presents many of his female characters as intelligent, sometimes superior to some of the males of the plays. One of the best examples is Portia from *The Merchant of Venice.* Poor Bassanio gets himself into such a dilemma by borrowing money from his good friend Antonio that he does not know how to help when Shylock demands either the return of his money or a pound of flesh. Portia, however, does not dally long in her decision to act. When Bassanio tells her that Antonio's ships are still out to sea, she sees no real problem for once she and Bassanio are married, there will be more than enough money to pay the bond: "You shall have gold / To pay the petty debt twenty times over" (III.ii.309). After Bassanio insists that he must go to his friend who is in such trouble, Portia decides to take her maid Nerissa to a monastery to wait for news. Then she tells her servant Balthasar to take a letter to her cousin, Doctor Bellario. The letter is a request to borrow her cousin's legal robes; Portia is scheming to challenge this Shylock in court in order to legally release Antonio from his bond. We see this come to fruition in Act IV of the play when, amid the accusations by Shylock against Antonio, in walks Nerissa, dressed as a lawyer's clerk, and Portia, dressed as a Doctor of Laws. When Portia fails to arouse sympathy within Shylock with her famous "quality of mercy" speech, she uses her wit, citing an old law of Venice stating that one causing the blood of a Christian to be shed would be required to forfeit lands and goods unto the state. With such gloomy prospects, Shylock is willing to accept payment of thrice the bond. To show her full knowledge of law, Portia then cites a further statute: anyone who even threatens the life of a Venetian citizen will have to deliver to the citizen offended one half his goods and then give the other half to the state. Shylock is caught up in his own web of vengeance, Antonio is free of the bond, and now Portia and Bassanio can marry. Albeit in man's attire, a woman has saved the day and made not only her husband happy, but also herself, and in turn she has satisfied the desires of all Venetian Christians for vengeance on a hateful Jew. Now this intelligence demonstrated by Portia flies in the face of those inherited attitudes of the Renaissance, inherited from centuries of the superior feelings of males toward females which suggested that women were not only weak physically and fickle sexually, but also lacking in wit. In demonstrating such intelligence, however, Portia has found it necessary to dress herself in a man's robe and pretend to be what she is not. Some of Shakespeare's women are much more forthright in demonstrating their superiority of knowledge without pretense and, in doing so, in fact become true mentors for men.

One such female is Paulina in *The Winter's Tale.* We first meet Paulina when she goes to prison to inquire after the health of her lady, Hermione. Most students of the play remember that Leontes, without much justification, has accused his wife, Hermione, of adultery; she has defended herself and argued her love for Leontes. Nevertheless, he has still cast her away into prison. Paulina is not allowed to speak to

Hermione, but she does converse with a waiting woman, Emilia, and discovers that Hermione has borne a daughter a little earlier than planned, one likely to live. Paulina's first plan is not to condemn the king or teach him any lessons, but merely to show him the little baby and hope that the sight will affect his senses. Convincing Emilia that it must be done, and convincing the jailer that no danger will come to him, Paulina takes the child and makes her way to Leontes.

Her husband Antigonus attempts to stop her from entering, but she is determined and proceeds to show the child to Leontes, who immediately lashes out in rage at her: "Out! / A mankind witch! Hence with her, out o' door! / A most intelligencing bawd!" (II.iii.66–68).

At this point, Paulina becomes the mentor. She calls the king a madman and rebukes her own husband for following the wishes of the king. When the king in turn chastises Antigonus for allowing his wife to rule him, Paulina replies that the king should allow his wife, Hermione, to rule him and then he would, in reason, accept his own children. She follows by identifying the king's opinions as rotten "as ever oak or stone was sound" (II.iii.89). Leontes continues to believe that the new daughter is not his, but Paulina presents a very convincing argument that the child must be legitimate by nature when one observes the physical likeness of the child and Leontes, the "copy of the father, eye, nose, lip. / The trick of's frown, his forehead, nay, the valley, / The pretty dimples of his chin and cheek, his smiles, / The very mould and frame of hand, nail, finger" (II.iii.100–3). When Leontes threatens to burn Paulina at the stake, she argues that she cares not, for it is the heretic that sets the fire. She calls him a tyrant whose cruel usage of his queen makes him shameful to the world. After Paulina exits, Leontes in another rage instructs Antigonus to take the child and burn it, else he will dash out its brains and then punish Antigonus for not following orders. Convinced that Antigonus set his wife Paulina on him about the child, Leontes entertains various lords who swear that such is not true. Following their convincing argument, Leontes determines that Antigonus will take the child into the wilds and allow chance either to destroy or to rescue it.

Paulina has exited, but she has not completed her task as mentor. During the trial of Hermione, a Delphic Oracle is read that declares her innocence as well as the innocence of Polixenes and Camillo. Then a servant appears to announce to the king that his son Mamillius is dead. Suddenly, Leontes repents and Hermione swoons, giving Paulina the room she needs to teach Leontes the real lessons of life. She re-enters and lashes out at Leontes with her severest language, claiming that his tyrannies and his jealousies have caused him to run stark mad, denouncing also the horrible crimes of the man, crimes that have led to the death of Hermione, whom he has just forgiven. All this, we learn later, is of course staged, for Hermione is not dead at all: she has listened to Paulina, who serves as her mentor, also.

Paulina has planned that Leontes will suffer grief for many years before he sees his wife again. Leontes has accepted the counsel of Paulina, realizing that he is guilty of all the crimes of which she has accused him: "Good Paulina, / Who hast the memory of Hermione, / I know, in honour, O, that ever I / Might have look'd upon my queen's full eyes" (III.i.49–53).

In the conversation following, Leontes swears to Paulina that he will never take another wife, and Paulina presses the point asking him to swear never to marry but with her permission. Leontes swears and goes even further to say that he will allow Paulina to choose that new queen should he ever find a candidate suitable. With an element of foreshadowing, Paulina remarks that such will never be until Leontes' first queen, Hermione, is in her breath again. This all fits in neatly with the finding of Perdita, the lost child, by the shepherd, and with the love affair of Perdita and Florizel. When the discovery scene occurs and Perdita is found to be the lost princess, happiness is partially restored to Leontes. He would be happier, of course, should he have his wife back. Paulina has been hiding her all these sixteen years. When Paulina offers to show Leontes a statue of the likeness of Hermione, Leontes accepts, only to discover it is the real woman. Paulina has kept Hermione in hiding until the time was ripe and Leontes has proven his word. Paulina thus has become the female mentor for both Leontes and Hermione. The circumstances prove that no one in the play is suited for the role of mentor except Paulina.

All's Well That Ends Well is another controversial Shakespeare play. For one thing, there is no way of dating the play, unless it was originally titled *Love's Labour Won*, a title appearing in Francis Meres' famous list, but a play not extant. There is no quarto version; therefore the folio copies must suffice; but they are full of inconsistencies of character, movement, and plot. Also curious, Shakespeare seemingly allows a gentlewoman, someone of a lower social order, to become the female protagonist of the play, but in doing so he allows Helena to become yet another female mentor to men.

The source is not difficult to identify, for it is the story of Giletta of Narbon in William Painter's *Palace of Pleasure,* a translation providing the ninth novel of the third day in Boccaccio's *Decameron.* In this story, Giletta (in Shakespeare, Helena) is rich, and living with her kin, though sought after by many husbands. After she has cured the king of a terrible illness, she chooses Beltramo (in Shakespeare, Bertram) for her husband. Beltramo rejects her, so she goes to his estate, repairs it, receives accolades from her subjects, and waits for her husband. When he does not choose to return, she devotes herself to pilgrimages and devotion. The story does include, but rather vaguely, an action of a helper (Diana in Shakespeare's play) who assists Giletta to deceive Beltramo into thinking he has slept with another when it is really his wife. Giletta remains in Florence until she has borne twins; then she takes the babies to Rossiglione, where Beltramo has secluded himself. She confronts him at a great banquet he is giving, where he admits the children must be his. Beltramo then accepts his wife.

In Shakespeare's play, Helena is not rich but is the orphaned daughter of a medical doctor, now taking refuge in the palace of the Countess of Rossillion, Bertram's mother. Helena is more on a social level with Parolles, a braggart soldier who plans to attend Bertram as Bertram journeys to the court of the king in Paris. In the opening scene, Helena matches her wits with those of Parolles in a discussion of virginity. Parolles is arguing a strong *carpe diem* position, which has its own pragmatic validity, but he finds Helena is able to match his word play quite well. He departs, saying that he will return the perfect courtier. Helena's first action in the play thus *appears* to be that of mentor to a male who desires to return reformed so that he can properly instruct her in male/female relationships. Parolles says:

> *I will return perfect courtier; in the which, my instruction shall serve to naturalize thee, so thou wilt be capable of a courtier's counsel and understand what advice shall thrust upon thee; else thou diest in thine unthankfulness and thine ignorance makes thee away.* (I.i.221–26)

Helena's real task, however, is to serve as mentor to Bertram, a very selfish young man who is addicted to surface values. Helena has fallen in love with him while staying in the household of his mother, but he is unaware. The Countess, nonetheless, is not unaware; she has noticed during the many weeks of Helena's stay that Helena has been touched by her son. When the Countess inquires of Helena her true feelings, Helena is at first embarrassed, remembering that she is of a lower social order and that such a love would not be appropriate. After some countering, Helena finally confesses:

> *Then, I confess,*
> *Here on my knee, before high heaven and you,*
> *That before you, and next unto high heaven,*
> *I love your son.* (I.iii.197–200)

The Countess accepts Helena's true feelings well. When the Countess asks if Helena had not expressed a desire to venture to Paris, Helena responds that she does want to try one of her father's remedies on the ailing king. The Countess realizes that should Helena be successful in Paris, she might win not only the king's love but her son's love as well; thus she gives Helena her blessing, and Helena sets off, with attendants offered by the countess, for the city.

At first the king refuses the remedy offered, but eventually he gives in, agreeing that if Helena can cure him, he will give her any man in the kingdom for her husband. She is successful, and the king keeps his word; she is granted Bertram, but he rejects her because she is of lowly birth. The king is quite angry with Bertram and his argument:

> *I know her well;*
> *She had her breeding at my father's charge.*
> *A poor physician's daughter my wife! Disdain*
> *Rather corrupt me ever.* (II.iii.120–23)

The king finds this difficult to believe, arguing in an extremely democratic manner that being good is sufficient without a good name. When the king decrees that the wedding will take place regardless of Bertram's attitude, Bertram decides to go to war to escape marriage. He sends his "wife" a letter of regret, and she decides to leave on a pilgrimage.

As time passes, we find Bertram quite a woman chaser, and he is now after a girl named Diana, who desires him not. When Helena shows up in Florence, she learns of Bertram's exploits. She decides she will play a trick on Bertram and teach him about love. The letter Bertram sent stated that he would never marry her unless she could get his ring off his finger and produce his child. With Diana's help, Helena is able to get Bertram into a compromising situation. Bertram's insistence that Diana sleep with him allows her to give in, with the promise that he will give her his ring. Thus Bertram spends a dark night with a woman he thinks to be Diana.

Time passes and Bertram returns to Rossillion, assuming that the information that his "wife" is dead is accurate. There he weeps along with the rest of the family in honor of her. When the king arrives he notices a ring upon Bertram's hand that once belonged to Helena. Helena had exchanged a ring with him, as Diana had promised him. In the darkness, he had not known that it was Helena or her ring. Suddenly, Diana enters the scene and accuses Bertram of the dark evening alliance earlier. By this time the king is terribly confused, but Helena, not dead at all, enters to explain the situation. Now that she has explained it all satisfactorily, Bertram willingly accepts her defense and takes her happily as his willing bride. When the king agrees to pay Helena's dowry, all is joy, and the play ends happily, with a man having been taught a valuable lesson by intelligent Helena.

Therefore, we see in such examples that Shakespeare uses female nobility to explore various stereotypes of the times. One could cite Hotspur's Kate and Brutus' Portia as other examples. What seems so very interesting is that Shakespeare also does this with women of a lower socioeconomic order than Helena.

A good play to begin with is *Othello*. Emilia, handmaiden to Desdemona, acts as more than a utilitarian character. She gives a passionate lecture on sexuality and the rights of wives. This occurs when in Act IV, scene iii, Othello tells his wife to dismiss Emilia and get to bed. In talking with Emilia at this point, Desdemona sounds like the dutiful, understanding wife of the ancient world. Emilia at first responds in a functional way, noting simply that she has laid the sheets on the bed as commanded. As their conversation about Othello continues, Emilia responds with an earthy tone, alluding to Othello's sexual desirability. Such comments lead to a dialogue on sexual morality.

This dialogue highlights Desdemona's naive purity and Emilia's pragmatic earthiness:

Des.	*Dost thou in conscience think—tell me, Emilia—* *That there be women do abuse their husbands* *In such gross kind?*
Emil.	*There be some such, no question.*
Des.	*Wouldst thou do such a deed for all the world?*
Emil.	*Why, would not you?*
Des.	*No, by this heavenly light!*
Emil.	*Nor I neither by this heavenly light;* *I might do't as well i' th' dark.* (IV.iii.61–69)

Such a dialogue is important to notice because, ironically, the subject is one neither could openly discuss with a husband. Desdemona finds it difficult to believe that women who abuse their husbands could exist. Emilia's response is a beautifully effective defense of wives against male double standards:

But I do think it is their husbands' faults
If wives do fall. Say that they slack their duties,
And pour our treasures into foreign laps;
Or else break out in peevish jealousies,
Throwing restraint upon us; or say they strike us,
Or scant our former having in despite:
Why, we have galls; and though we have some grace,
Yet have we some revenge. Let husbands know
Their wives have sense like them; they see, and smell,
And have their palates both for sweet and sour,
As husbands have. What is it that they do
When they engage us for others? Is it sport?
I think it is. And doth affection breed it?
I think it doth. Is't frailty that thus errs?
It is so too. And have not we affections,
Desires for sport, and frailty, as men have?
Then let them use us well; else let them know,
The ills we do, their ills instruct us so. (86–103)

Although Bianca's role is the smallest of the three female roles in *Othello*, it is nevertheless an important one. As a woman conventionally scorned, desired, and used by men, she underlines the theme of female abuse. Although she is a difficult woman, a bit crusty, her speech shows that she is also passionate and honest:

What? keep a week away? seven days and nights?
Eightscore eight hours? and lovers' absent hours,
More tedious than the dial eightscore times?
O weary reck'ning! (III.iv.173–76)

Bianca is really the most outspoken person in the play. Her forwardness and honesty counteract the egocentrism of both Othello and Cassio. She brings fear into men through her forwardness:

> *Let the devil and his dam haunt you! What did you mean by that same handkerchief you gave me even now? I was a fine fool to take it. I must take out the work?—A likely piece of work, that you should find it in your chamber, and not know who left it there! This is some minx's token, and I must take out the work? There; give it your hobby horse: wheresoever you had it, I'll take out no work on't.* (IV.i.153–61)

As Eamon Grennan says in his article on the female voices in *Othello*, "In its own way her speech is a moral reality that brings into sharper focus the moral deficiencies of the world that would condemn hers" (283).

Another example of wisdom is that of Hostess Quickly in *Henry IV, Part I.* Agnes MacKenzie in her 1924 book *The Women in Shakespeare's Plays* assesses Mrs. Quickly's character similarly with that of Dickens: "Mrs. Quickly walks on earth, or rather scuttles over it like the beetles of a London basement. There is a touch of tenderness in her" (103). One should add that there is also a touch of wisdom in this tavern hostess.

Her opening lines do not convey her sense of intellect, but when in Act III she argues with Falstaff, her lodger, over money, she wins, and to win in wit over him is an achievement:

Fal. *How now, Dame Partlet the hen! Have you inquired yet who picked my pocket?*

Host. *Why, Sir John, what do you think, Sir John? Do you think I keep thieves in my house? I have searched, I have inquired, so has my husband, man by man, boy by boy, servant by servant. The tithe of a hair was never lost in my house before.*

Fal. *Ye lie, hostess: Bardolph was shaved and lost many a hair; and I'll be sworn my pocket was picked. Go to, you are a woman, go.*

> *Host.* *Who, I? no; I defy thee: God's light, I was never called so in mine own house before.*
>
> *Fal.* *Go to, I know you well enough.*
>
> *Host.* *No, Sir John; you do not know me, Sir John. I know you, Sir John: you owe me money, Sir John; and now you pick a quarrel to beguile me of it: I bought you a dozen of shirts to your back.* (III.iii.60–76)

Falstaff seems to be getting back at her with his wit, but then the Prince arrives. The Hostess now sees her chance to get law and order on her side, so she takes advantage of it with an amazing torrent of outraged virtue. When she tells the Prince that Falstaff has said he has spoken vilely of him, Falstaff denies it. She affirms that it is the truth, or "There's neither faith, truth, nor womanhood in me else" (III.iii.125).

Falstaff retorts by calling her a "thing" and telling her to go. This allows the Hostess to reply with virtuous sentiment:

> *Host.* *I am no thing to thank God on, I would thou shouldst know it; I am an honest man's wife: and, setting thy knighthood aside, thou art a knave to call me so.* (III.iii.134–36)

Finally, Falstaff must own up to the theft himself, although he tries to have the last word by ironically "forgiving" the Hostess and then telling her to go make his breakfast, love her husband, look to her servants, and cherish her guests; and then assures her that he is pacified.

An interesting example of group intelligence and wit can be seen in the last act of *The Merry Wives of Windsor*, a play cast with the lower social order in mind. The subject matter is love among gentlemen, but it is love mixed heavily with the action of commoners and servants.

In Act V, the wives trick Falstaff into entering Windsor Park at midnight. There, with their unusual magical powers, powers not usually associated with the merchant class, they transform him into a deer, which action, he realizes, makes him an "ass," not unlike the transformation of Bottom in *A Midsummer Night's Dream*:

> *Fal.* *The Windsor bell hath struck twelve; the minute draws on. Now, the hot-blooded gods assist me! Remember, Jove, thou wast a bull for thy Europa; love set on thy horns. O powerful love! that, in some respects, makes a beast a man, in some other, a man a beast. You were also, Jupiter,*

a swan for the love of Leda. O omnipotent Love! how near the god drew to the complexion of a goose! A fault done first in the form of a beast. O Jove, a beastly fault! And then another fault in the semblance of a fowl; think on't Jove; a foul fault! When gods have hot backs, what shall poor men do? For me, I am here a Windsor stag; and the fattest, I think, i'the forest. Send me a cool rut-time, Jove, or who can blame me to piss my tallow? Who comes here? my doe?

Mrs. F. *Sir John! are thou there, my deer? my male deer?*

Fal. *My doe with the black scut! Let the sky rain potatoes; let it thunder to the tune of Green Sleeves, hail kissing comfits and snow eringoes; let there come a tempest of provocation. I will shelter me here.*

Mrs. F. *Mistress Page is come with me, sweetheart.*

Fal. *Divide me like a bribed buck, each a haunch: I will keep my sides to myself, my shoulders for the fellow of this walk, and my horns I bequeath your husbands. Am I a woodman, ha? Speak I like Herne the hunter? Why, now is Cupid a child of conscience; he makes restitution. As I am a true spirit, welcome!* (V.v.1–26)

Though some treat this passage as underlining the lustful nature of Falstaff and the episode's parallel with the Greek story of Actaeon and Diana, another focus here is that of giving ordinary folk—especially the women Hostess Quickly and Mistresses Page and Ford—magical powers. Nancy Cotton sees this image as one of witchcraft, but actually the text uses the term "fairy" and the entire scene suggests Titania and Oberon's fairy land rather than the world of Macbeth (Cotton 321–26). Although both worlds involve the use of supernatural powers, there is a negative connotation placed upon witchcraft. True, fairies could be mischievous, like Puck in *A Midsummer Night's Dream,* but were usually not considered evil.

Add to this what appears to have been Shakespeare's purpose in writing and producing the play, and the use of magic for commoners seems especially important. David Bevington reminds us that according to eighteenth-century tradition, Shakespeare composed the play at the behest of the Queen. John Dennis asserted in 1702 that Elizabeth I was "so eager to see it acted, that she commanded it to be finished in fourteen days." Nicolas Rowe added in 1709 that the Queen, having been so pleased with Falstaff in the *Henry IV* plays, wished to see him in love. Such accounts must be regarded with caution, but they do point to a passage of courtly flattery unlike anything we find elsewhere in Shakespeare. A fairy blessing bestowed on Windsor Castle in Act V is obviously intended to celebrate the famous Order of the Garter. Mistress Quickly orders her fairy charges to sing nightly "Like to the Garter's compass, in a ring," and to tend carefully the several chairs of order—those decorated stalls in the Chapel of St. George belonging to the illustrious lords who made up the Order of the Garter. Every such installment receives her blessing, along with each knight's coat, crest, and blazon.

For Shakespeare to present his women in such light as shown in the various plays cited goes against the grain of stereotyped attitudes of the times toward females, attitudes inherited from centuries of the past. But Shakespeare, in his subtle, intelligent manner, is clever enough to see through the superficiality of society's norms. It behooves us in contemporary times to note this ability.

Works Cited

Bevington, David. *The Merry Wives of Windsor,* "Introduction." *The Complete Works of Shakespeare.* Glenview, Illinois: Scott Foresman and Company, 1980.

Clark, Mary Cowden. *The Girlhood of Shakespeare's Heroines..* 3 vols. London: W.H. Smith, 1850.

Cotton, Nancy. "Castrating (W)itches: Impotence and Magic in *The Merry Wives of Windsor.*" *Shakespeare Quarterly* 38 (Autumn 1987): 321–26.

Dash, Irene. *Wooing, Wedding and Power: Women in Shakespeare's Plays.* New York: Columbia University Press, 1981.

Dusinberre, Juliet. *Shakespeare and the Nature of Women.* New York: Barnes and Noble, and London: Macmillan, 1975.

Grennan, Eamon. "The Women's Voices in *Othello*: Speech, Song, Silence." *Shakespeare Quarterly* 38 (Autumn 1987): 283.

Jameson, Anna. *Shakespeare's Heroines: Characteristics of Women.* London: Saunders and Otley, 1832.

MacKenzie, Agnes Mure. *The Women in Shakespeare's Plays*. (London: William Heinemann Ltd., 1924.

Painter, William. *Palace of Pleasure.* New York: Dover, 1966 .

Pitt, Angela. *Shakespeare's Women.* New York: Barnes and Noble, 1981.

Shakespeare, William. *The Merchant of Venice,* reprinted in *The Complete Works of Shakespeare,* ed. David Bevington (Glenview, Illinois: Scott, Foresman and Company, 1980). All citations of the plays, i.e., *All's Well That Ends Well; Henry IV, Part I; The Merry Wives of Windsor; Othello;* and *The Winter's Tale* , are from this edition.

Some Aspects of Shylock's Jewish Nature

by Shirley Marney

Much of the critical attention given to *The Merchant of Venice* centers on Shylock: his character, his role, his relation ship with other characters and with society. Even more important, a question arises about Shylock's relationship with his author, bringing forth the issue of whether Shakespeare, and *The Merchant of Venice*, are anti-Semitic. Critical answers range from those of E. E. Stoll, who sees Shakespeare's Shylock as both an utter villain and a comic butt,[1] and D. M. Cohen, who sees *The Merchant of Venice* as "profoundly and cruelly anti-Semitic,"[2] to the ideas of J. B. Thompson, who sees Shylock as "deeply religious"[3] and defends him by attacking the play's Christian characters as "obsessive" about money and "both materialistic and snobbish."[4] Thompson, far from seeing the play as anti-Semitic, sees it as breaking down traditional stereotypes of good Christian and bad Jew.

Most of the questions about Shylock revolve around his Jewishness. Is he hero or villain? Is his conduct justified by the prejudice he has met? Is the play a plea for tolerance or a vehicle of racism? Where do the author's sympathies lie? Is this a play that can be produced today without distortion of one kind or another? While absolute answers to these questions cannot be found, I believe that one can approach them more closely by examining an issue not thoroughly explored by critics: the nature of Shylock's Jewish character—specifically, the basis of his character and his characterization.

Many of the discussions of Shylock use the term *Jew* as though there were only one meaning to it, one definition, one set of connotations.

The issue might be clearer if this notion were true, but the word carries many meanings, and the critics themselves use it with variations. While we cannot hope to pinpoint what exactly the term meant to Shakespeare or to the Elizabethan audience, we can look at what the play itself tells us about the term as used with regard to Shylock.

The first possibility is that Shakespeare made Shylock the stereotyped stage Jew, a villainous and barely human monster, the essential alien, clad outlandishly, with a long, hooked nose and red hair. This Jew is the embodiment of unspeakable evil and, in a comedy, the buffoon and, at the same time, the comic butt. This view is, I think, essentially that which Stoll has of Shakespeare's drawing of Shylock.

From this perspective, externally and internally, in appearance and in character, Shylock is the outsider, the threat to community, the scapegoat. Taking the matter of appearance first, Stoll says,

> *It is highly probable, moreover, that Shylock wore the red hair and beard ... as well as the bottle nose of Barabas. So Judas was made up of old; and in their immemorial orange tawny, high-crowned hats and "Jewish gabardines" ... the very looks provoked derision.*[5]

However, despite tradition, there is no reason supported by the play itself to suppose that Shylock looked different from anyone else. If he had a long nose, a yellow hat, and a myriad of sartorial appurtenances that the Middle Ages or the Renaissance associated with Jewry, then Portia's question, "Which is the merchant here and which the Jew?" (IV.i.174)[6] is nonsense. M. Channing Linthicum's scholarly inquiry into Shylock's "gabardine" indicates that it was likely a garment that could have been worn by anyone, even the nobility, and which therefore had no connotation of racial or social status. This study reproduces an illustration of a Jewish merchant of Padua, printed in 1563, and concludes that "the Jew of the City could not be distinguished from Christians of his profession (merchant) except that possibly his cap was yellow."[7] Cecil Roth agrees, stating that "their clothing was not outlandish, however, nor did they wear special garments," though he believes that the hat should probably be crimson.[8] Other internal evidence from the play consists of the negative fact that no one, not even the irreverent Launcelot or the sharp-tongued Gratiano, mentions anything about Shylock's clothing or appearance.

Going on to the character of the stage Jew, one argument supporting Shylock as the stereotypical embodiment of evil is that one primary source for Shylock is Christopher Marlowe's Barabas, title character from *The Jew of Malta.* Barabas is the incarnation of power, of revenge, and of evil. The fact that he is a Jew gives form to the incidents of the play, but it does not really shape his character. He is despoiled of his riches by the city rulers because he is a Jew (Jews being historically a source of revenue for rulers); he is further enraged by his daughter's conversion to Christianity; and his vengeance takes the form of murderous attacks against figures that are markedly Christian (poisoning the whole convent, strangling a friar). However, his genesis is not from racial roots, but from the revenge tragedy and from studies of ruthless Machiavellian power figures, such as Marlowe's own Tamburlaine. As N. W. Bawcutt points out, Barabas is "utterly self-centered" and "has no links with any group of human beings (not even his fellow Jews)."[9]

Barabas is wronged by Christians, but his wrath against them comes because they oppose him, not because of their religious beliefs (or his). His enmities and his (temporary) alliances are based on Machiavellian, not theological or racial, grounds. His only real allegiance is to himself and the power goals he has set. Shylock may owe the fact that he is a Jew to Marlowe's source, though the office of usurer carries this implication. Shylock may owe to Barabas his daughter and his confused loyalty to the daughter and the money. He may also be so indebted for his deadly thirst for revenge, though the form of the revenge is quite different in Shakespeare's comedy than in Marlowe's tragedy. Despite these influences, Shylock is not Barabas translated to Venice and the comedic mode.

A major difference is that Shylock is much more of a Jew than is Barabas. Barabas' Jewishness is important to the plot of *The Jew of Malta* because it gives the city's rulers a reason for despoiling him and thus provides him with a reason for seeking revenge. It gives him a second grudge in his daughter's conversion, which leads him to a wonderful exhibition of villainy. But any wrong done to a vengeful character can lead to revenge. The poisoning of an entire convent may be particularly shocking to an audience, but it is only a variation of the blood feud, the convent being the offender's new family. Race and religion are not really important to Barabas' character or his values.

On the other hand, Shylock's Jewishness is central to his character. While it is not, as will be seen, the cause of his villainy, it is clearly central to his thinking. Shylock's Jewishness is the pivotal point of his concepts about himself; he hardly opens his mouth without mentioning it. It colors his view of the world and everyone in the world. (It also colors everyone's view of him, but to a much lesser degree; they think of him primarily as a usurer.) Shylock's conviction that anyone who opposes him does so out of racial prejudice is of questionable validity. Norman Nathan says Shylock makes "an unwarranted assumption that the merchant is anti-Semitic."[10] Certainly Shylock makes an association between his "sacred nation" and his "thrift."

It probably should be noted that another aspect of the traditional stereotype of the Jew that Shakespeare did not borrow is racial prejudice. While Shylock may think of his Jewish nature as racial, it is not. Jessica, daughter of Shylock and Leah, does not share his nature. No one raises the least objection to her marriage with Lorenzo (except Shylock, of course), which they would do if race were the issue. Jessica, on her own initiative, separated herself from her father's character ("manners")[11] though not from his blood (race).

Shylock's Jewish nature, in the eyes of the other characters in the play, and I think in the eyes of the play's author, lies in his religion. And it is in the religious realm that the true sources for Shylock's character lie. He is a Biblical Jew. *The Merchant of Venice* is filled with Biblical allusions, references, and parallels, and none is more predominant or more important than the story of Jacob. First introduced by Shylock (I.iii.72), the story permeates the play and tells, I think, much of at least one aspect of the character of Shylock, and of Jessica, and of the relationship between Shylock and Antonio as well. Through Jacob's veins ran the bluest blood in Jewish history: he was a patriarch, the grandson of Abraham, and the progenitor, through his twelve sons, of the nation which bears his name (both Israel and Jacob are names of Jacob, and both are used repeatedly throughout the Bible to refer to the Jews as a nation). He was the father of kings.

However, as an individual, Jacob was hardly a role model. The fact that Shylock chooses him for one is telling. Jacob was a master of taking personal and profitable advantage of a situation. His first recorded action was the sharp trade he made with his elder twin

Esau of a "mess of pottage" for Esau's birthright, exchanging a bowl of lentil soup for the rights of inheritance of the firstborn.[12] While one may condemn Esau as stupid (a quality he manifests elsewhere as well) and as despiteful of his birthright, Jacob's action does not speak well of his character, particularly as it was unnecessary; he already had divine assurance of the birthright.[13] His next action was still more culpable. Inspired and abetted by his mother Rebekah (whom Shylock styles as "wise" [I.iii.74]), Jacob deceived his blind father through lies and a disguise, and stole Esau's blessing, which amounted to primacy in the family. His actions earned his objectives, but they also earned him his brother's hatred and a twenty-year exile. In the incident cited by Shylock, the rods were Jacob's means of pre-natal influence, which he used to profit over his equally shifty Uncle Laban, Rebekah's brother. His methods, sharp, at least as Laban and Laban's sons saw them, forced another hasty exit, and only divine intervention allowed Jacob to escape with his life, his property, and his family. He returned to his promised inheritance and his family position only after a theophany and after humbling himself to the prosperous Esau. Against this background (tradition, nation, prototype), Shylock appears.

Shylock's choice of Jacob as his justification, his prototype (a role belonging much more to Jacob than to Barabas), is significant. Jewish history, as presented in the Bible, is filled with heroic, larger-than-life-sized, characters—Abraham, Moses, and David, to name only three—and while many of these characters have flaws, they retain qualities which make them worthy role models. Shylock's choice, Jacob, has as his most notable characteristic a sort of acquisitive shrewdness. He knows how to gain wealth by whatever means are available.

Shakespeare's characterization of Shylock as a Biblical Jew consists not only of his Old Testament association with Jacob, but with the New Testament portrayal of the Jewish religion as degenerate (Harold Bloom calls this a New Testament "slander.")[14] Shylock is presented as a Pharisee. The Pharisees were the conservative element of the Jewish nation, separatists, upholding the Law in its most rigorous form (though from a Christian perspective a Law shorn of its spiritual content and consisting only of its formal element), and the most determined opponents of Christ and the Christians. Shakespeare establishes Shylock's character as a Pharisee from his first scene. When Antonio appears, Shylock calls him a "fawning publican" (I.ii.42), a puzzling reference (since Antonio neither fawns nor is a publican, i.e., tax collector or quisling) unless the epithet is associated with the parable of the Pharisee and the publican.[15] In this parable the Pharisee stands on his own righteousness while the publican depends on God's mercy. The despised publican, of course, is justified before God.

Shylock's later reference to Abraham, "our holy Abram" (I.iii.73), is also an association with the Pharisees, who rebutted Christ's message by asserting that Abraham was their father.[16] Christ rejected their claim and told them that the devil was their father.[17] In establishing Shylock as a Pharisee, Shakespeare has set the foundation for Shylock's self-righteousness and his fanatical dependence on the law as well as his association with the devil.

Further association of Shylock with the devil is made by Antonio ("The Devil can cite Scripture for his own purpose" in I.iii.99); by Launcelot ("the Jew is the very Devil incarnal" in II.ii.28); by Salanio ("Let me say 'Amen' betimes, lest the devil cross my prayer, for here he comes in the likeness of a Jew" in III.i.22); and by Bassanio ("And curb this cruel devil of his will" in IV.i.218). This wholesale condemnation comes not because Shylock is a member of a hated race, but because his actions and attitudes mark Satan as his spiritual leader. The Pharisees of the New Testament were not condemned for their adherence to their Jewish heritage, but because they had deviated from it.

Shylock's association with the devil becomes even more interesting because not only is he a murderer, in intent at least, and has deviated from or rejected the truth, but he plays an additional role Biblically associated with Satan. He is the accuser of the elect. The Pharisees were the accusers of Christ, but the connection goes beyond that. The Greek word *diabolos,* translated into English as "devil," means "accuser" or "calumniator." The Bible calls Satan "the accuser of our brethren … , which accused them before our God day and night."[18] And so Shylock functions in the play, accusing Antonio of slandering the Jews, accusing Christians in general of seeking vengeance and of committing the inhumanity of slavery. The deceit behind his accusations is not that Christians or Christian society is not guilty of these faults, but that he is justified because the Christians are flawed.

The height of Shylock's self-justification comes when he is challenged by the Duke: "How shalt thou hope for mercy, rendering none?" (IV.i.88). Shylock's smug "What judgement shall I dread, doing no wrong?" (IV.i.89) is a denial of the Christian position that all have sinned under the Law. This denial is what Russell Astley calls the "Jewish heresy," the belief that adhering to the Law negates the need for Grace.[19] Portia, speaking for Christianity, is well aware of the "naughty world" around her:

Though justice be thy plea, consider this,
That in the course of justice none of us
Should see salvation. We do pray for mercy .... (IV.i.198–200)

Portia knows that the Christians of her world are not perfect, that they have need of mercy. She knows as well that Shylock also has need of mercy, a fact that he refuses to accept as he stands Pharisaically on his own righteousness. Portia uses the Law, which is without mercy, to attempt to bring Shylock to a realization of his own need. Her concept of the law and the position of all men under it is thoroughly in accordance with that of St. Paul, who teaches that both Jew and Gentile have sinned, that "by the Lawe commeth the knowledge of sinne,"[21] that justification comes only through Grace,[22] and that race or religious creed has nothing to do with the issue.[23] Shylock is condemned, not because he is a Jew, but because he has claimed justification under the law, which can only condemn. Portia has been criticized for rendering Shylock only what the law allows him, but orthodox Christian doctrine contends that only by learning the true purpose of the law can Shylock learn of Grace. He asks for law; she gives him law that he may see a need for mercy.

Barbara Lewalski suggests that the courtroom scene is an allegorical confrontation of the Old Law and the New,[24] but while this may be in part true, the tension between Justice and Mercy existed in the Old Testament as well as in the New. In the Holy of Holies of the Tabernacle was the Mercy Seat.

Nor is Shylock's guilt simply positional. Even if one ignores the reinterpretation of the law in the New Testament and looks only at the Law as given to the Jews, Shylock falls far short. First, he seeks revenge, though the Law forbids it to individuals, regarding vengeance as belonging to the Lord.[25] Second, he refuses to show mercy, which is a Christian virtue, but which is also required by the Old Testament and the lack of which is condemned.[26] Also, while the *lex talionis* is sometimes regarded as justifying bloodthirsty revenge, it actually does the opposite. In the contexts in which it appears in the Bible,[27] it limits the reparation for an injury to the damage caused by the injury. Shylock is entitled, under the Law, to the return of his ducats, not to Antonio's life. If Antonio has insulted him, he may return Antonio's insults (which he does), but no more. His reference to the synagogue (III.i.135) indicates that he thinks of it not as a place of worship and prayer, but as the scene of further plotting with Tubal. The vows he makes are not promises to God, but promises of revenge. The obsessions which Shylock has with money and with vengeance indicate the degradation not only of his character, but also of his religion. He sees his own loss of daughter and ducats as the ultimate crime against the Jewish nation. It is significant that what he values in his ancestry is not the spiritual heritage, but the ability to make money grow.

Less significant to the eternal order of things, but still a factor in a comedy that condemns Shylock for his Pharisaic behavior and attitudes, is that he is what might be termed a Jewish Puritan. Puritans were seen by high church types (and perhaps by much of the general populace) in Shakespeare's day as the Christian equivalent of the Pharisees. Doctrinally, the Puritans might be sound, but they were seen as being self-righteous and often hypocritical. They were also generally regarded as opposing anything frivolous or extravagant in life. Shylock's sober house and his opposition to music, feasting, and revelry exactly fit this characterization. Jessica calls her home "tedious" (II.iii.3) and mourns the departure of Launcelot, who gave it its only hint of merriment. It is in this guise of Puritan that Shylock is most surely to become a comic butt; heavy fathers in comedies were ever deceived by daughters running away with lovers, and such fathers have ever contributed involuntary dowries to the absconding couples.

Paul A. Cantor says that Shylock is "destroyed by his own virtues," referring most specifically to his "devotion to the law,"[28] and I believe the point is well taken, but the virtues of this Pharisaical Puritan are a cold, lifeless legalism that leads to self-righteousness and hypocrisy. Shylock's self-righteousness appears from his first scene when he clearly regards Antonio's financial dealings as inferior to his own careful thrift and when he sneers at the dietary habits of the Christians who have invited him to dinner. Religious slurs and prejudices are clearly not the sole prerogative of his enemies. For those who see his offer of friendship as a ruse designed to bring Antonio into his power, this action is arrant hypocrisy, as is his "merry bond."

Conniving opportunist, Pharisee, Puritan—Shylock is far from being the stereotypical comic butt or the equally stereotyped Jewish monster. He is a villain and is forcibly dragged back into the circle of community, as is befitting in a comedy. His fault is not his race (Christian Jews are acceptable). I agree with Moody E. Prior: "The antipathy to Shylock as a Jew in the play does not arise from some property in him which is regarded as by nature ineradicable from him and his descendants."[29] Shylock's faults come partly from his religion (his version of the Jewish religion), but mostly from his character. Shylock's religion fails him because it trusts to legalism and self-righteousness rather than mercy, and his character fails because he values money (solitary, sober thrift) over community (risk-taking for the sake of others). Shylock's choices, and his failure, are mirrored in the casket scenes and in the choices of Portia's hopeful suitors. Morocco chooses gold, wealth ("what all men desire"), Aragon chooses silver, justice ("as much as he deserves"), and the prodigal Bassanio wins the prize with lead, he being willing to "give and hazard all he hath" for the chance of winning his love. Shylock is defeated, not because he is Jewish, but because, even within that framework, he makes wrong choices. He chooses Jacob; he chooses to demand justice over mercy. His choices come not because he is Jewish, but because he is the kind of man he is. He could have chosen Abraham, who was generous, and he could have chosen mercy.

What kind of Jew is Shylock? Is Shakespeare's portrait the result of prejudice? There are still no absolute answers, but in my opinion, Shylock's destruction in the play comes because of his character, not race, not even really religion. A Jewish Puritan no more has to be a Puritan because he is a Jew than a Christian Puritan has to be because he is a Christian. The purpose of comedy is the establishment, or re-establishment, of community. The "happy ending" requires that Shylock be brought into the fold, willy-nilly, but that fold is one of life and love rather than the sterility of barren metal.

Notes

[1] E. E. Stoll, "Shylock," reprinted in *William Shakespeare: Comedies and Romances,* ed. Harold Bloom (New York: Chelsea,1986).

[2] D. M. Cohen, "The Jew and Shylock," *Shakespeare Quarterly* 31(1980):53.

[3] J. B. Thompson, "The Modification of Stereotypes in *The Merchant of Venice,*" *ESA* 26 (1983): 3.

[4] Thompson 1.

[5] Stoll, 13.

[6] Quotations from *The Merchant of Venice* here and elsewhere in this study are from the G. B. Harrison edition.

[7] M. Channing Linthicum, "My Jewish Gaberdine," *PMLA* 43 (1928):761.

[8] Cecil Roth, "The Background of Shylock." *Review of English Studies* 9 (1933): 154.

[9] N. W. Bawcutt, ed., Introduction. *The Jew of Malta* (Manchester: Manchester UP, 1978) 22.

[10] Norman Nathan, "Shylock, Jacob, and God's Judgment," *Shakespeare Quarterly* 1 (1950): 256.

[11] MV

[12] Gen. 25.34.

[13] Gen. 25.25.

[14] Harold Bloom, Introduction, *William Shakespeare: Comedies and Romances* (New York: Chelsea, 1986) 3.

[15] Luke 18.10–14.

[16] John 8.39.

[17] John 8.44.

[18] Rev. 12.10. The Geneva Bible.

[19] Russell Astley, "Through a Looking Glass, Darkly: Judging the Hazards of *The Merchant of Venice,*" *Ariel* 10 (1979): 21.

[20] Rom. 3.9.

[21] Rom. 3.20.

[22] Rom. 3.24.

[23] Gal. 3.28.

[24] Barbara Lewalski, "Biblical Allusion and Allegory in *The Merchant of Venice,*" *Shakespeare Quarterly* 13 (1972):334.

[25] Deut. 32.35; Psa. 94.1; confirmed by Rom. 12.13.

26 Prov. 11.17, 21, 26; Hosea 4.1; Micah 6.8.

27 Exod. 21.24; Lev. 24.20.

28 Paul A. Cantor, "Religion and the Limits of Community in *The Merchant of Venice*," *Soundings* 70 (1987): 254.

29 Moody E. Pryor, "Which is the Jew that Shakespeare Drew?" *The American Scholar* 50 (1981): 484.

The Work Laws, Prosperity, and *The Tempest*

by Esther M. Gloe

When Shakespeare wrote his history plays, he tried to make the action fit the era in which it took place; thus, for instance, he did not portray King Lear as a Christian. In his book, *The Dream of Prospero*, D. G. James observed that *The Tempest* seems to contain some kind of allegory, but he did not describe one in particular.[1] Robert Pierce observed that *The Tempest* seems to inspire a good deal of confusion and multiple interpretation.[2] To arrive at a clearer interpretation of Shakespeare's *The Tempest*, a romance, not a history, one must look at the notable events which were taking place in England at about the time when Shakespeare was writing the play. He wrote for the people of his own time, and what they were interested in was what he wanted to put on the stage. To relate our thinking to *The Tempest*, we should remember the passage in England of a new work law in 1610 and the publication between 1595 and 1609 of writings about the new Virginia Colony. These things were the talk of the town in 1611 when *The Tempest* was first performed.

During Shakespeare's last years of writing for the stage, a near obsession with work was a major motivating force among the people of England. Work had begun to replace war as man's main occupation.[3] The Puritans, who were growing in number and influence at that time, studied Calvin, and in the final, 1559, edition of his *Institutes of the Christian Religion*, Calvin had included a defense of all labor as being what God wanted man to do. It reads as follows: "Therefore, lest through our stupidity and rashness everything be turned topsy-turvy, he (God) has appointed duties for every man in his particular way of life."[4] By 1601, Calvin's defense of labor had become in Arthur Dent's *The Plaine Man's Pathway to Heaven* an insistence that all people must work at something.[5] The Puritans wanted everyone to work. Shakespeare did not write to please the Puritans, nor were most of his audience Puritans, but those thoughts about work were often uppermost in their minds because the Puritans were increasingly becoming a force to be reckoned with. Besides that, Shakespeare, himself, was a hard worker, and the idea that all people should work probably appealed to him.

Though the first English work law had been enacted in 1349, the most inclusive one was enacted by the Parliament of 1609–10, and it required compliance by September 29, 1611.[6] Therefore, it is not surprising to find that *The Tempest*, written in 1611, served its author as a vehicle for looking at attitudes toward work, a popular topic of the times. Other factors back up such an interpretation of the play. An analysis of the probable meanings of the names of the characters in the play and an analysis of the setting of the play help substantiate interpreting *The Tempest* as an allegory about work and people's attitudes toward work. So also does the probable date of the play's first public performance, sometime not long before September 29, 1611.

Many scholars have shown connections between *The Tempest* and events which took place shortly before the play was written. Sir Adolphus Ward was one of the first to link the speeches of Gonzalo (II.i.147–64) to Montaigne's essay, "Of the Cannibals." He showed that Gonzalo's description of an ideal commonwealth was taken from that essay, and that Montaigne's essay was published in English for the first time in 1603.[7] D. G. James calls Chapter IV of his own book "The New World," and in it he lists the writings about the colony in Virginia which had been published by 1609, some of which were probably read by that avid reader, Shakespeare.[8] Ernest Gohn mentioned the play's many references to the words "now" and "at this time," and these words do appear six times in Act I, Scene ii (37–38, 169, 239, 249, 403, 432), the first scene after the shipwreck, and also they appear three times in the Epilogue (1, 3, 13). However, Gohn connected this to his own idea that the play was the only one of Shakespeare's plays which adheres to the classical unity of time.[9] It is more appropriate to say that they are related to Shakespeare's knowledge of contemporary events.

The main event which influenced Shakespeare's thinking was the passage of the Statute of 1610 which tightened up the work laws. By removing exemption clauses, the new law made it possible to send to the workhouse persons who had sufficient income that they did not have to work, if they were arrested for disorderly conduct.[10] It specifically exempted from the law only one type of worker: the shipwrecked sailor.[11] To see the connection between the play and the law, one should review the development of the laws dealing with work and idleness, consider the attitudes of the people toward work, look at statements already made by critics about the relationship of the play to sloth, look at the meanings of the names of the characters, look at the pun inherent in the play's structure, and study the play itself to see what evidence there is in the actions of the characters to connect the play to the new law enacted in 1609–10.

From the time of the Black Death on, the English government passed laws dealing with problems of employment. The Statute of 1349 was passed after the Black Death had left the country with a shortage of workers, especially farm workers. The law regulated wages because lords had begun to adjust wages so as to get such free labor as was available.[12] Next, in 1353, the Statute of Staple became law, creating one of the earliest monopolies, the "Staple Towns," which had a monopoly on selling certain things.[13] Then, in 1406, the first Statute of Apprentices was enacted to keep farm workers on the land by specifying that a man had to own land worth at least 20s in order to apprentice his children.[14] (Workers had been moving to the cities to get higher wages.)

After 1500, the picture changed, but it was not until 1563 that the famous Statute of Apprentices was passed, attempting to organize all laws regulating education, work, vagabondage, begging, and relief of the poor. It tried to regulate work in such a way that there would be enough workers in the menial jobs, the skilled trades, and the educated professions. It recommended that gentlemen's younger sons be apprenticed, because if they were not, they would either become vagabonds or get their fathers to hoard money so as to leave them a living.[15]

The poor laws—also called work laws or vagabond laws—of 1572, 1576, 1597–1601, and 1604 attempted to correct the problems which arose after the passage of the law of 1563. The law of 1572 said that vagabonds should be put to work, that scholars must have a license from the university in order to beg, and that actors should be considered vagabonds unless they could produce a patron to vouch for their employed state.[16] The law of 1576 said that towns should supply themselves with flax or wool for the poor to work with, and it provided that the poor who would not work should be sent to a workhouse which each town should build.[17] The laws of 1597–1601 said that poor children should be apprenticed to the humble trades so that they would not become beggars, and that scholars should no longer be allowed to beg.[18] The law of 1604 said that poor children should be apprenticed at government expense.[19] Sidney and Beatrice Webb suggest that the work laws tended to be aimed at whatever group of idle people had most recently offended.[20]

The law of 1609–10, which especially influenced *The Tempest,* was brought into being by the excessive number of idle gentlemen living in London. Their presence was so noticeable in the London of 1608 that the Lord Chancellor, speaking to judges and justices of the realm in the Star Chamber, rebuked the crowd of "new and young knights who come in their braveries and stand there like an idol to be gazed upon, and do nothing."[21] Some of these young men were justices of the peace, who were supposed to be sending the poor and vagabonds to workhouses but were not doing so. It is best to note here also that it was the lack of work skills of similar minor gentry which largely caused the failure of the original Virginia Colony. Captain John Smith wrote: "A hundred good workmen would be worth a thousand such gallants."[22] D. G. James presents evidence that Shakespeare knew the colony organizer, Strachey, and was familiar with the colony in general.[23]

The people of that time also considered monopolies granted to courtiers to be a factor which increased unemployment—by depriving men of jobs. Therefore, the feeling against monopolies went hand in hand with the feeling aroused by the unemployment of the poor and the unwillingness to work displayed by the minor gentry. The situation was brought to a head by several law cases in which Sir Edward Coke played an important part. In 1599–1600, Coke's case, *Davenant vs Hurdis,* introduced the idea that all monopolies were "against the common right."[24] Then, in 1603, the famous case, *Darcy vs Allen,* contested the twenty-one year patent in playing cards by challenging the Crown's right to give patents at all.[25] Next, in 1605, *Allen vs Gerard* was filed by Allen, a London merchant, who resisted Darcy's patent, saying that all men should labor because it was the law of God, and firmly stated that all monopolies should be void.[26] Voiding monopolies did not become the law until 1623.

Resentment of monopolies, of idle knights who loafed around London, and of people who were simply too lazy to work reached a peak in 1609. Therefore, the Parliament of 1609–10 wrote a law which tightened up the system.[27] It made specific provision for building workhouses, putting people in them, and fining justices of the peace who did not enforce the law. The new law removed from the books the exemption "though they have lawful means to live by" regarding gentlemen. It classified as a beggar anyone who was released from jail after a disorderly conduct charge—if he had to find someone to pay his bills there.[28] Thus, a gentleman could no longer casually spend a night in the counter and expect to be released as he had in the past. The law retained only one exemption, the one which permitted shipwrecked sailors to beg. The people of London were familiar with these laws. As an audience, they would have realized that the sailors in *The Tempest* suddenly disappeared from the play after the shipwreck, and they would have thought of the new law.

The laws were related to the Puritan ideas about work and calling. Idleness had never been praised by the church; the old morality plays sometimes showed idleness as the father of all vices. The Puritans started with that idea and the belief that work kept people busy and thus less likely to fall for temptations. They moved on to the idea that work had value in itself; it was a "calling" from God. Part of this was inspired by a misunderstanding of the teachings of Calvin. According to Max Weber, this special emphasis developed in England because of the use of Cranmer's version of the Bible.[29] To Calvin, the "calling" was the call to election—salvation by faith—sent from God to a person. It made no difference what the occupation was so long as the believer worked at it in such a way that he was obeying the call to faith. Calvin's work reads: "From this will arise also a singular consolation: that no task will be so sordid and base, provided you obey your calling in it, that it will not shine and be reckoned very precious in God's sight."[30]

It took a while for this respect for honest labor to become a matter of interest to English Puritans; however, by 1601, this had been accomplished. One of the most famous religious writings of that time was Arthur Dent's *The Plaine Man's Pathway to Heaven*, which says on the subject of work: "God doth allow none to live idely; but all both great and small are to be imployed one way or another, either for the benefits of the Church or Common-wealth ... or for the good of Towns and Parishes, and those amongst whom they do converse ... or for one good use or another."[31]

The Puritans saw idleness not only as the origin of poverty and unemployment, but even more as neglect of the gifts which God had given to people. They encouraged the passage of work laws and vagabond laws. William Haller traces Puritan thinking through sermons and other writings of the preachers who served as lecturers in the churches, especially in the prominent churches of London, in his book, *The Rise of Puritanism: 1570–1643.*[32] The questions of work and calling were kept alive by the Puritan lecturers. Since the Puritans could not take over the established church and modify it, they devoted their energy to attempts to perfect the faith and lives of the people directly, and this was the work of the Puritan lecturers. The first lectureship was established in 1559 at St. Antholin's parish church.[33] By 1582, at least thirty of the 116 parishes of London had lecturers, half of them known to be Puritans; by 1629, there were 107. People flocked to hear such famous speakers as Greenham, Rogers, and Perkins.[34]

The Puritan lecturers actually competed for audiences with the theaters because the lectures were usually given on workdays. Therefore, the lecturers tended to dramatize the things which they taught rather than deal in abstract doctrine. Since the topic of work was of such general interest, it is not surprising to find that it was also dramatized by other playwrights. One play which deals with the topic of work is Dekker's *If This Be Not a Good Play the Devil Is in It* (1612), in which the king angers his courtiers by deciding to leave only Saturday free for pleasure.[35] A well-known play is Middleton's *A Chaste Maid in Cheapside* (1613), in which characters of all classes live without doing any work at all.[36] The best known other play is probably Ben Jonson's *Bartholomew Fair* (1614).[37] The Induction to Jonson's *Bartholomew Fair* connects it to *The Tempest* in lines 129–32, which reads: "He is loth to make Nature afraid in his plays, like those that beget

TALES, TEMPESTS, and such like drolleries, to mix his head with other men's heels."[38] Obviously, Shakespeare's play was a success, or his friends would not have tried their hands at the same subject.

In addition, the Puritans constantly criticized the players for doing no observable work. Since the Puritans criticized them, the players were always eager to catch the Puritans or members of other religious groups doing something which could be considered as useless as theater work, and a playwright as smart as Shakespeare would prefer to make a point that the Puritans would agree with.

The Tempest was written immediately after the 1609–10 law was passed. The new law stressed that every shire, borough, and town was to construct a workhouse where idle people could be put to such work as spinning and weaving.[39] The deadline for accomplishing this was the Feast of St. Michael the Archangel, September 29, 1611. Justices of the peace who did not have a workhouse operating by that date were to be fined five pounds. The law made it possible for a gentleman caught in disorderly conduct to be confined to the workhouse. Only shipwrecked sailors were still permitted to beg, and then they were permitted only to beg their way home from the seacoast.[40]

A. L. Rowse says that in *The Tempest* Shakespeare is writing about one of his favorite topics, the work of a ruler.[41] This time, Shakespeare uses his ability to create puns as devices in order to portray the differences between workers who do their jobs well and those who do not. The play may be interpreted as an allegory of the problems of work in the world, the work of a ruler, and whether or not it is right to expect the gentility to work at something. The play was written between the time the law was passed in 1610 and the time it became effective in 1611. *The Tempest* begins with a scene which portrays sailors facing shipwreck.

Several critics have already observed that *The Tempest* appears to deal with the questions of idleness and sloth. Harry Epstein says that Prospero lost his dukedom because he preferred occult studies to his duties as ruler.[42] William Rockett says that while perfecting his mind, Prospero neglected worldly ends and is finally engaged in the labor of building a new life out of the ruins of his past.[43] Leo Marx says that Prospero's control at the time of the action of the play is based on hard work, study, and scholarly self-discipline.[44] Harry Berger says that Prospero has learned that he can rule only by rigid control when confronted with such subjects as Caliban and Antonio.[45] It is apparent that Prospero is a person who is learning to handle work properly.

The speeches of Gonzalo furnish a contrast to this concentration of Prospero on work and control. Gonzalo refers to ideas found in Montaigne's essay, "Of the Cannibals," which describes a society in which no formal economic system or political system is found. However, such ideas were also common among some of the more radical religious sects, such as the Anabaptists, as described by White.[46] Another such sect, the Family of Love, was satirized by Thomas Middleton in 1607 in his play called *The Family of Love.*[47] Therefore, most of the people in the audience would have recognized the ideas. To the scornful Sebastian and Antonio, Gonzalo describes the kingdom he would set up on the island if he were governing it:

> *Gon.* *Letters should not be known; riches, poverty,*
> *And use of service, none; contract, succession,*
> *Bourn, bound of land, tilth, vineyard, none;*
> *No use of metal, corn, or wine, or oil;*
> *No occupation; all men idle, all;*
> *And women too, but innocent and pure;*
> *No sovereignity—*
>
> *Seb.* *Yet he would be king on't.*
>
> *Ant.* *The latter end of his commonwealth forgets the beginning.*
>
> *Gon.* *All things in common nature should produce*
> *Without sweat or indeavor.* (II.i.151–61)

Rockett maintains that Gonzalo shows only the negative aspect of such a society, idleness, and adds that Gonzalo does not seem to see the redeeming efficacy of work.[48] However, Dean Ebner suggests that by having Gonzalo state Montaigne's ideas, Shakespeare sets up the action so as to deliver a gentle refutation of Montaigne's views.[49] There is something to be said for both ideas. Gonzalo may be stating information as much as displaying his own personal desires. However, Gonzalo does not demonstrate much respect for work when he comments on the work of the sailors in the opening scene:

Alon. *Good boatswain, have care. Where's the master? Play the men.*

Boat. *I pray now, keep below.*

Ant. *Where is your master, boatswain?*

Boat. *Do you not hear him? You mar our labor; keep your cabins, you do assist the storm.*

Gonz. *Nay, good, be patient.*

Boat. *When the sea is. Hence! What cares these roarers for the name of King? To cabin!*

Gonz. *Good, yet remember whom thou hast aboard.*

Boat. *None that I more love than myself. You are a counsilor; if you can command these elements to silence, and work the peace of the present, we will not hand a rope more; use your authority.* (I.i.11–25)

In relation to Montaigne's ideas about primitive societies, one should observe that contact with the primitive Caliban does not cure Stephano and Trinculo of their wicked, civilized ways. In addition, as Maxwell Luria points out, when Sebastian calls himself "standing water," he is claiming a constitutional idleness. The passage is so short that it is easy to miss:

Seb *Well, I am standing water.*

Ant. *I'll teach you how to flow.*

Seb. *Do so. To ebb*
Hereditary sloth instructs me. (II.i.221–23)

Luria uses these lines to connect the message of the play to criticism of sloth.[50]

Daniel C. Boughner points out that the shipwreck can symbolize the dissolution of a society which the play separates into three parts: Prospero, Miranda, and Ferdinand; Stephano, Trinculo, and Caliban; and Gonzalo, Antonio, Alonso, and Sebastian. He connects this to the classic structure of the play, not to the specific character portrayals.[51] However, when they are grouped in this way, the characters can be related to their attitudes toward work. First, one should note that the actual manual laborers in the play are sailors, and they are eliminated from the action in all but the first and last scenes.

They are shipwrecked sailors, such as were exempted from the work law. These sailors do not beg their way home, however; they are magically put to sleep.

Ernest Law and Irwin Smith discuss an additional factor, the idea that the direction of the play as it now exists is distorted by the fact that after its first performance, the original masque (IV.i.60–156) was replaced by one more suitable for the play's presentation at the wedding celebration for the Princess Elizabeth.[52] The new masque is one which portrays the goddesses of love and fertility. The original masque was a dance of the rabble.[53] The term "rabble" was not necessarily pejorative at that time, but simply meant "crowd" or "vagabonds." What this dance of the rabble would have been like, we do not know, but it could have been movements of the vagabonds being required to work. They could have been the same "Strange Shapes" which helped Ariel present and remove the banquet for Alonso, Antonio, Gonzalo, and Sebastian. In Act IV, scene i, line 37, Prospero tells Ariel: "Go bring the rabble ...." But in the modern version of the play, Iris, Ceres, and Juno come on stage at line 60.

Further indication that the intended meaning of the play is one related to work comes from the pun involved in the play's basic structure. As Eric Partridge's etymological dictionary shows, the word *travel,* "journey," is the same word in English as the word *travail,* "work."[54] Skeat adds that the two words were spelled alike at the time Shakespeare lived.[55] In his *Variorum* edition of *The Tempest,* Furness mentions that Shakespeare possibly picked the names Alonso, Antonio, Gonzalo, and Sebastian from Eden's *History of Travaille,* which was published in 1577.[56] *Travaille* is the French spelling for "work," but Eden's *History of Travaille* is about travel. Only in English do the words *travel* (v) and *travail* (n) come from the same root.[57] Molly Mahood refers to a similar pun on travel and travail in Sonnet 63, and calls this pun one of Shakespeare's "favorite portmanteau words, packed with two meanings."[58] She adds that the word is also implicitly used in this way in *Richard II.* There, in Act I, Scene 3, John of Gaunt says to Bolingbroke about his exile: "Call it a travel that thou takest for pleasure" (II.ii.262). Bolingbroke is not satisfied with this and in his reply turns the word travel into travail:

> *Bol.* *Must I not serve a long apprenticehood*
> *To foreign passages, and in the end,*
> *Having my freedom, boast of nothing else*
> *But that I was a journeyman to grief?* (I.iii.271–74)

This pun was apparently one of Shakespeare's favorites. In *The Tempest,* it is obvious that only Caliban is a native of the island where the action takes place, and Caliban does not like work. All the other characters are travellers, and they do work at something. *The Tempest* is an extended metaphor, built on a pun, in which all of the travellers are thought of as workers.

The first group of characters is the heroine, Miranda, her father, Prospero, and the hero, Ferdinand. The second group is the Duke's own servant-managers, Sebastian and Trinculo. The third group is members of the upper class, or hangers-on at court. From this third group, Gonzalo, speaking lines from Montaigne's essay in quasi-jest, provides Shakespeare with a popular statement from outside his own land which shows what the opposite of hardworking Puritan society might be, without making it necessary for him to appear to cast aspersions on any sect which was then popular.

The names of the three main characters—Prospero, Miranda, and Ferdinand—carry work-related meanings from ancient times.[59] In the context of travel, Prospero is he who facilitates travel; Miranda is an admirable voyager; and Ferdinand makes a hard journey. But in the context of work, the names change their functions. Prospero is he who controls prosperity, and Miranda is, as Patrick Grant has suggested, Charity.[60] Prospero seems to make this connection early in the play when he says to Miranda:

> *Pro.* *Oh, a cherubim*
> *Thou wast that did preserve me. Thou didst smile,*
> *… A noble Neopolitan, Gonzalo,*
> *Out of his charity, who being then appointed*
> *Master of this design, did give us …* (I.ii.152–53,161–63)

Ferdinand is a hard worker, a man who carries logs for a whole day. The names *Prospero* and *Ferdinand* are standard; their meanings were well known in Shakespeare's time, and it would be unlikely for him to assign them some other significance. In this context, it is logical that "Prosperity" be the father of "Charity," just as Prospero is the father of Miranda.

In the group of servant managers (Trinculo, Stephano, and also Caliban) the names can best be interpreted in terms of their sound just as travail and travel are. Trinculo is a trinkerman, which according to *The Oxford English Dictionary* was a well-known term between 1538 and 1720, when it was used to describe a person who fished with a trink, a type of fixed fishing net. The word trink was used as a legal term from the 14th century to the 17th century because the use of the trink in the Thames was regulated by law.[61] Trinculo catches Caliban by using a "gabardine," a type of rain cloak, for a net. He makes fun of Caliban's fish-like qualities when he first meets him, saying: "A man or fish? Dead or alive?"(II.ii.25). Trinculo is a jester who "fishes" for applause, a common meaning for the word "fishing" already then; he tries to joke about the fish-man.[62] Trinculo works at jesting even when he is not on the job. As a worker, Stephano is almost a stevedore who carries casks of liquor off the boat. The word "stevedore" did not exist then; however, the word from which it evolved was in common use. The word was *steven*, the Scotch dialect term for the forecastle of a ship, the place where things are stowed.[63] In the forecastle, Stephano, the butler, stored the wine. As monopolist of wine, he interferes with the ability of others to do their work. The name for Caliban, the island's servant, can come from several sources. Caliban can come from the names of the land Caribe (or *Calibe* as it was sometimes spelled), or it can come from Calibia in North Africa.[64] The name *Caliban* can also simply be "Cannibal" with the syllables reversed and the letters mixed up. It makes no difference whether Caliban is Carribean or African or simply a cannibal, he is a primitive. Not until after 1890 did critics think that Shakespeare might be portraying American Indians as Caliban.[65] Englishmen of 1609 had met only a very few people from any spot in the New World. They had not met the five northeastern tribes of the Iroquois Nation—who had made peace among themselves and turned from warfare to agriculture. They had not met the Five Civilized Tribes of the southeastern United States—who were such hard-working farmers that they proved to be too much competition for the settlers, who asked that they be moved away, which resulted in their being sent west to Oklahoma. Caliban is simply Montaigne's cannibal or "noble savage" because Shakespeare is making fun of the idea that one could study such savages as a source of enlightenment. The name Caliban was devised to make the audience think of Montaigne's essay. Shakespeare portrays Caliban as having no interest at all in work. Thus, this group of servant managers is made up of one who refuses to do work, one who jokes about work, and one who handles his work—a monopoly—badly. On the other hand, Ariel serves to contrast with the non-Christian Caliban. His name means "Lion of God" in Hebrew, but its sound connects it with the air and the heavens in a way that would be obvious even if Shakespeare and his audience knew no Hebrew.[66] Ariel is a hard worker, and he eventually earns his freedom by faithful service.

To unravel the meanings of the names of the remaining group, the gentry, one can relate the characters to their parallels in the lower-class group. The nobles all misuse work. The name *Sebastian* is related to the Sebestan plum, a fruit used in the preparation of medicine.[67] Sebastian is also lazy, as is Caliban, and the two are roughly equivalent in that they plot to overthrow a ruler but do not succeed. Sebastian is thus a person who misuses his talents. The name *Alonso* means one who is eager to go to battle.[68] The name *Alonso* also sounds like "along." Alonso went along with Antonio's deposing of Prospero (I.ii.121–35). Thus, Alonso is a person who misuses his power. He is also the out-of-date man who still dreams of war rather than work. The name *Antonio* has no known meaning; however, the name was that given by swineherds to the smallest pig of the litter, the one designated as a gift to the church because it had the least value.[69] In addition, a sound-alike word is *autonomasia*, the rhetorical term which signifies the substitution of epithets for names.[70] Therefore, this name, *Antonio*, may be a tipoff to what is going on: epithets are being used as names in the play. Antonio is a parallel type to the monopolist, Stephano. Antonio is the ultimate monopolist. He takes over an entire dukedom by substituting himself for the ruler. Thus, he is one who misuses trust placed in him. The name *Gonzalo* sounds like "counselor," and he is called one. However, the other characters constantly laugh at his ideas, and the Italian word *gonzo* means "fool."[71] The "foolish" ideas of Montaigne are spoken by Gonzalo, the counselor, for the amusement of the rest of the party. Thus, he is a person who misuses his position by playing the fool. All of these gentlemen misuse work: they do the wrong things, they go along with wrong ideas, they hold a monopoly of power, they praise foolish notions. *Adrian* and *Francisco* are names which simply label their owners as aliens: Adrian is from Hadria and Francisco is from France.

The significance of the names and of the pun on travel and travail are reinforced by two other things. Many interpreters believe that the performers in the play should pronounce the name of the city as if it were "my lan'" and not in the Italian way.[72] Thus, the story appears to be about "my land." The title of the play contains another pun, this one related to the time of the action. The word *tempestive* meant "timely" (*tempus*) during that era, so the play, *The Tempest,* was dealing with a timely question.[73] If the drama had been written by a modern television script writer, it probably would have been titled "*Our Times.*"

Though critics once thought that the play was not presented until 1613, it is now generally agreed that it was performed before James I on November 1, 1611.[74] This would have been just one month after September 29, 1611, the date by which the work law was supposed to have been enforced. It is generally agreed that Shakespeare's later plays were never performed for the first time before the King. According to Ernest Law, this play is not mentioned by Dr. Simon Forman as being one he saw at the Globe in April and May of 1611.[75] *The Tempest* was publicly performed sometime between June 1 and September 29 of 1611, about the time when the country first was feeling the effects of the new law. The play was Shakespeare's own piece of work done to pay for his share of the new theater, Blackfriars, according to Ernest Law.[76] Shakespeare got his workhouse constructed and operating before the deadline.

With the names and the puns explained, the plot has a definite meaning. Prosperity and Charity were driven out of "my land" when the people in power neglected work for occult studies and seemed to expect magic to get the work done. "My land" was taken over by monopolists who drove Prosperity and Charity out. But as Prosperity has a good deal of power and control, Prosperity sought out the evil workers and arranged to bring them to justice here in a little workhouse-island-theater. A future ruler, who is a hard worker, could keep such a situation from happening again. He would, however, have not only to work hard, but also to work for love of Charity. Charity will help him rule wisely—and work should always provide for charity. Magic, however, is a method which only operates successfully on faraway islands inhabited by primitives. Nothing helpful can be learned by studying the ways of non-working primitives. Work done in the right way can solve many of the problems of "my land." Such an interpretation is allegorical. The play was not aimed at any particular individual. It is a picture of what can happen in a society when both the members of the ruling class and the members of the other classes do not carry out their duties properly.

A look at the actions of the characters substantiates this interpretation. In the first group of characters, Prospero has learned that he cannot neglect his own work for a life of study; to be a ruler, he must rule. It is easy to see a resemblance between Prospero and James I, who had a great intellectual curiosity and a tendency to get directly involved in projects. As George Unwin remarked, James I did not have much ability to tell a good project from a bad one.[77] However, it would be unwise to make a direct comparison between James I and Prospero because Prospero is merely a duke, and a great many of the gentry and nobility of that time were involved in the economy as holders of patents of monopoly. Many gentry needed to learn Prospero's lesson: stick to the job. And those who were trying to colonize Virginia particularly needed to learn how to do some useful work. And people in control should learn that their control—if wisely used—makes them purveyors of prosperity.

Ferdinand is a hero fit for an era in which work has replaced warfare as the road to glory.[78] Ferdinand knows that he must work for his true love, so he carries logs. Caliban despised the chore and complains about it:

> *Cal.* *Here comes a spirit of his, and to torment me*
> *For bringing wood in slowly. I'll fall flat,*
> *Perchance he will not mind me.* (II.ii.15–17)

Ferdinand carries logs to prove his nobility and to prove he is a worthy husband for the fair Miranda. No job is too sordid or base for the hard worker, as Calvin said. But Ferdinand does not enjoy the work itself. He says:

> *Fer.* *There be some sports are painful, and their labour*
> *Delight in them sets off; some kinds of baseness*
> *Are nobly undergone, and most poor matters*
> *Point to rich ends. This my mean task*
> *Would be as heavy to me as odious, but*
> *The mistress which I serve quickens what's dead*
> *And makes my labours pleasures.* (III.i.1–7)

He knows the value of work in relation to charity. Without the one, the other would not be possible.

Carrying logs was long a task used in plays, but probably more importantly, it was also one that the gentry balked at doing in Virginia, where lumber was supposed to be the first export from the colony back to England, where lumber was badly needed.[79] In the morality play, *The Disobedient Child* (1553), Disobedient Child will not go to school, will not work, and will not become a soldier. His father finally gives him his inheritance early, and he and his young wife soon spend all of it. When the money is gone, his wife finds him a job—carrying wood. Late in the play, the father philosophizes about it saying: "Some work to do it must be thy care."[80] Carrying logs was also the last task assigned to the young woman in Dekker's Griselda story, *Patient Grissill* (1599). In this play, Grissill's brother, a scholar, helped her with the other tasks but finally balked at carrying logs.[81] Grissill proved her true devotion to her noble husband by doing the work. Shakespeare picks this familiar task to make his point. Ferdinand proves his good intentions by carrying logs. The idea that everyone should work was not a new one, but in 1610 it became the law; and by the next time a crew was sent to Virginia, Captain Smith enforced the law there too by threatening with banishment to the wilderness those drones who would not work.[82]

In contrast, the noble Miranda does not carry logs. Her job is to be a good wife. She should succeed at it because she does try to help Ferdinand. She even offers to help carry logs:

> *Mir.* *If you'll sit down*
> *I'll bear your logs the while. Pray give me that,*
> *I'll carry it to the pile.*
>
> *Fer.* *No, precious creature,*
> *I had rather crack my sinews, break my back*
> *Than you should such dishonour undergo,*
> *While I sit lazy by.* (III.i.23–38)

Ferdinand declines her offer. In the context of allegory, charity is help freely given to those who are not able to work, and Patrick Grant's analysis of Miranda as representing the virtue of Charity seems to be a good one.[83] As long as Ferdinand is able to do the work, he will do it himself. Shakespeare does not show him liking the work, but he does seem to say that Ferdinand is not degraded by doing work for a good reason and that his behavior is more admirable than the behavior of those who must be driven to do work. In addition, charity is to be admired, but it must not be substituted for work.

In the group of lowly characters, we find the opposite end of the spectrum of workers. Stephano, the forecastle man, is the duke's butler, something more than a servant. The butler was originally the importer of wines. Trinculo, the fisherman, is the court jester, who fishes for applause. Neither of these takes work very seriously. Their main occupation in the play is to make sure that they salvage some liquor from the wreck so that they can continue in their usual occupation, which seems to be "drunken and disorderly conduct" of the kind that in 1611 would send not only them but even a London gentleman to the workhouse. These two fail to reform early in their stay on the island, and they even seduce into their company the local primitive, Caliban. These characters are used by Shakespeare to refute the idea that the easy life of the primitive would do away with the problems which seem to be created by civilization. The fool and the drunk carry their problems with them and pass them along to Caliban, who is happy to join their group.

The middle group of characters—Alonso, Antonio, Sebastian, and Gonzalo—show attitudes toward work which are varied and ambiguous. Alonso has been working at his proper job, ruling Naples. His big mistake seems to have been being too eager to go along with the plans of Antonio, the usurper of the throne of Milan. Antonio has also worked hard; he did the job Prospero gave him to do, ruling Milan. However, he became too fond of the job and arranged to get rid of Prospero so that he, himself, could rule in his own name. In a way, he resembles what James I feared in his parliaments and law court: the desire of another to rule for him. However, in the context of work, Antonio is behaving like many of the monopolists or patent holders of England during the Tudor and Stuart periods, until 1623, when monopolies were at least theoretically forbidden. By 1600, most new patents or monopolies were held by courtiers. Other people resented the parceling out of both governmental administration and merchandising jobs to courtiers. Antonio is the ultimate monopolist. He takes over the entire government.

Sebastian is a different case; he frankly says that he is too lazy to work. In the very first scene, the Boatswain tells him that he should work rather than complain (I.i.45). Later, Antonio convinces him that he should exert himself enough to take over the kingdom when they get back to Naples. The idea did not occur to Alonso, who cannot see the connection between his approval of Antonio's deposing his own brother, Prospero, and the possibility that his own brother, Sebastian, might depose him. Neither Antonio nor Sebastian worries about how he will manage to get back to Naples to do any ruling. These characters are all more inclined to forceful takeover (equivalent to war) than they are to hard work, and they all work at the wrong things in the wrong ways. Hard work is not enough; it must be work which accomplishes something constructive. Symbolically, a free meal is offered to all of these men. Sebastian accepts it at face value; Alonso is dubious about it; Gonzalo, the gullible, talks them into trying it. Just as they are about to eat, Ariel removes it and reminds them of their past sins (III.iii). Only then does Gonzalo begin to suspect what the others are plotting.

The position of Gonzalo is very ambiguous. He has been a faithful counselor to both of the brothers who ruled Milan, and he managed to keep Prospero from being killed. However, he shows a lack of understanding of the value of work when he praises the glorious idleness that would abound in the land he would rule if he could, the land that seems to be Montaigne's land of the cannibals. The others do not take him seriously. Gonzalo's ineptness refutes not only Montaigne's ideas but also the ideas of writers of "utopias" and the ideas of religious sects which advocated communism or which dreamed of an ideal world. Shakespeare seems to think that such people would not make good counselors. Gonzalo should not be allowed to rule; he cannot keep Prospero (prosperity) in power. In addition, he does not seem to appreciate the hardworking sailors in the opening scene of the play, as was mentioned earlier. Gonzalo overvalues authority and does not realize that he is getting in the way of the working man who needs prosperity. The working man then reminds Gonzalo of his limitations. Gonzalo has good intentions, but he does not see the value of work itself; if he did, he could have explained it to Prospero some thirteen years earlier. He also does not seem to understand the problems of the about-to-be-shipwrecked sailors who must deal with the real world in a very special set of circumstances. Only the lawful duke or his heir can be a proper ruler, and the duke can only rule after he learns to stick to his ruling and leave magic alone. Later, however, Ariel does report that Gonzalo, as well as Adrian and Francisco, are much moved by the repentance of Antonio, Alonso, and Sebastian (V.i.11–15).

Considering the behavior of the characters, it becomes apparent that the play is a portrayal of the kinds of workers who are affected by the law which made it necessary for even a gentleman to do useful work. Prospero, the duke, failed when he neglected his own duties and took up those of a scholar. Gonzalo is a thinking counselor and a generous friend, but he cannot treat all classes fairly and he accepts any authority, regardless of whether or not it is properly in power. Sebastian knows that he is lazy, and he will act only when no real work is required. Antonio wants a job and takes it for himself by force. Antonio is the monopolist who runs the land that everyone else is too lazy or too distracted to take interest in. Ferdinand, a hard worker who appreciates the value of charity, is the man to be admired.

Such an interpretation of *The Tempest* should not surprise anyone who remembers that Shakespeare was a successful businessman. Gerald Eades Bentley has speculated that *The Tempest* was written especially for that new financial venture of the King's Men Company: the Blackfriars Theater. He suggests that this can help explain its difference from Shakespeare's earlier plays—it was written for indoor performance.[84] The new theater may also explain why Shakespeare had the question of work and workers on his own mind; he had to get the play finished and even write out the instructions for the actors as he was not often in London. William Armstrong mentions that Shakespeare "instructed" actors, or directed his own plays, as was reported in Sir William Davenant's Company after the Restoration.[85] Shakespeare had directed plays and helped run the company of which he was a principal stockholder. He knew what it took to rule the little realm where he was chief playwright, the Globe, and he knew the different types of jobs that different actors did. The types of workers found in this play could easily resemble the types of workers he met in the course of a day at the theater.

Only Ferdinand, the far traveller or hard worker, who is the future ruler of Milan and Naples ("my land" and some other place, maybe Virginia) appears to have the proper attitude toward work. He knows that he must work to prove that he is worthy of the lady, Charity. He does not pretend to like it; he simply does what needs to be done with a minimum of complaining.

With such a future ruler, the kingdom will be well cared for. With such a hero, the play can end happily. With such a message, the audience can feel secure, even if the next thing it hears may be a Puritan lecture.

It appears that Shakespeare made his statement on the value of work, particularly the work of the ruling class, in his romance, *The Tempest.* In so doing, he addressed a problem uppermost in the minds of the people of his land at that time, a time plagued with unemployment, monopolies, and a changing system of merchandising, as well as busy with exploration of a New World and speculation about such methods of organizing the economy as were described not only by Montaigne, but also by such sects as the Anabaptists, the Family of Love, and others. From this play, one can conclude that Shakespeare believed that learning or education of itself was not going to solve the problems of the country in some magical fashion. Neither would the problems be solved by utopian dreams. Only honest and earnest effort would prevail. The play spoke about a question timely in that era, and spoke in praise of each man doing his own job, doing it well, and expecting similar behavior from his fellows. When Caliban is freed, the play closes on a note of forgiveness for mistakes of the past. And in its Epilogue, it ends with a request for applause: "Let your indulgence set me free," says Prospero (Epi.20). Applause will set prosperity free for the playwright and his company by proving that they have produced another hit, and the audience can then produce its own prosperity by going back to work.

Notes

1 D. G. James, *The Dream of Prospero* (Oxford: Clarendon P, 1967) 2, 162.

2 Robert Pierce, "'Very Like a Whale': Scepticism and Seeing in *The Tempest*," *Shakespeare and History*, in *Shakespeare Survey* 38 (1985): ed. Stanley Wells (Cambridge: UP, 1985): 167.

3 Louis B. Wright, *Middle Class Culture in Elizabethan England* (Ithaca, NY: Cornell UP, 1935) 10.

4 Jean Calvin, *Institutes of the Christian Religion*, trans. and ann. Ford Lewis Battles, 12 vols. (Atlanta: John Knox P, 1975) I,iii, Ch. X, Sect. 6, 724.

5 Helen C. White, *English Devotional Literature: Prose* (Madison: University of Wisconsin P, 1931) 227. She quotes from Arthur Dent's *The Plaine Man's Pathway to Heaven* (1601) 191–92.

6 G. W. Prothero, ed., *Statutes and Other Constitutional Documents: 1558–1625*, 4th ed. (Oxford: Clarendon P, 1934) 272.

7 C. F. Tucker-Brooke, John William Cunliffe, and Henry Noble MacCracken, eds., *Shakespeare's Principal Plays*, 3rd. ed.(New York: Appleton, 1935) 900. They quote from Sir Adolphus Ward's earlier work.

8 James 75–88.

9 Ernest Gohn,"*The Tempest*: Theme and Structure," *English Studies* 45 (1962): 116–25.

10 David Little, *Religion, Order, and Law: A Study in Pre-Revolutionary England* (New York: Harper and Row, 1969) 202.

11 Little 212.

12 Theodore Plucknett, *A Concise History of the Common Law*, 5th ed.(London: Butterworth, 1955) 32.

13 A. K. R. Kiralfy, ed., *Potter's Outline of English Legal History*, 5th ed.(London: Street & Maxwell, 1958) 105.

14 Nicholas Orme, *English Schools in the Middle Ages* (London: Methuen & Co.,1973) 52.

15 George Unwin, *Industrial Organization in the Sixteenth and Seventeenth Centuries* (Oxford: Clarendon, 1940) 38.

16 Prothero 673.

17 Prothero 72.

18 Prothero 96–112.

19 Prothero 253.

20 Sidney and Beatrice Webb, *English Poor Law History: Part I: The Old Poor Law*, vol. 7 of *English Local Government* (1927 rpt. Hampden, CT: Archon Books, 1963) Part I.

21 Webbs 74–75.

22 Michael Kraus, *The United States to 1865*, vol.1 of *The University of Michigan History of the Modern World*, 16 vols. (Ann Arbor: The University of Michigan P, 1959–60) 40.

23 James 75–85.

24 Little 196.

25 Little 196.

26 Little 200.

27 Little 196–212.

28 Little 211–12; Prothero 101.

29 Max Weber, *The Protestant Ethic and the Spirit of Capitalism*, trans. Talcott Parsons (1904; rpt. New York: Scribners, 1958) 210 (note 3 to Ch. III); also relevant are 80, 113, 114, 178.

30 Calvin 725.

31 White 227.

32 William Haller, *The Rise of Puritanism: 1570–1643* (New York: Harper and Brothers, 1957).

33 Paul S. Seaver, *The Puritan Lectureships* (Stanford, CA: Stanford UP, 1970) 191–92.

34 Seaver 191–92.

35 Thomas Dekker, *Patient Grissill*, ed. Fredson Bowers, *The Dramatic Works of Thomas Dekker* (Cambridge: UP, 1958) vol. 1, 209–293.

36 Thomas Middleton, *A Chaste Maid in Cheapside*, ed. A. C. Swinburne, *Thomas Middleton: Unexpurgated Edition* (N.Y.: Scribners, 1887) vol. 1, 171–257.

37 Ben Jonson, *Bartholomew Fair*, ed. C. W. Herford and Percy and Evelyn Simpson, *Ben Jonson* (Oxford: Clarendon P, 1938) vol. 6, 9–141.

38 Harry Levin, *Shakespeare and the Revolution of the Times: Perspectives and Commentaries* (New York: Oxford UP, 1976) 212.

39 Prothero 271–72.

40 Prothero 271–72.

41 A. L. Rowse, *William Shakespeare: A Biography* (New York: Harper and Row, 1963) 236.

42 Harry Epstein, "The Divine Comedy of *The Tempest*," *Shakespeare Studies* 8 (1972): 282, 291.

43 William Rockett, "Labor and Virtue in *The Tempest*," *Shakespeare Quarterly* 24 (1973): 77, 78.

44 Leo Marx, "Shakespeare's American Fable," *The Massachusetts Review* 2 (1961): 57.

45 Harry Berger, "Miraculous Harp: A Reading of Shakespeare's *The Tempest*," *Shakespeare Studies* 5 (1969): 269.

46 White 38–42.

47 Middleton 25–110.

48 Rockett 72.

49 Dean Ebner, "*The Tempest*: Rebellion and the Ideal State," *Shakespeare Quarterly* 16 (1956): 168.

50 Maxwell Luria, "Standing Water and Sloth in *The Tempest*," *English Studies* 49 (1968): 328–29.

[51] Daniel C. Boughner, "Jonsonian Structure in *The Tempest*," *Shakespeare Quarterly* 21 (1970): 6.

[52] Ernest Law, "Shakespeare's *Tempest* as Originally Produced at Court" (London: Chatto & Windus, 1918) 24–26. This is a pamphlet for the Shakespeare Association, an address delivered in Feb. & Mar., 1918. Irwin Smith, "Ariel and the Masque in *The Tempest*," *Shakespeare Quarterly* 21 (1970): 221.

[53] Law 28.

[54] Eric Partridge, ed., *Origins: A Short Etymological Dictionary of Modern English* (New York: Macmillan, 1963) 734.

[55] Walter Skeat, *Etymological Dictionary of the English Language*, rev. ed. (Oxford: Clarendon, 1963) 661.

[56] Horace Howard Furness, ed., *The Tempest*, A New Variorum Edition of Shakespeare (Philadelphia: Lippincott, 1920) vol. 9, 4–5.

[57] Partridge 734.

[58] Molly Mahood, *Shakespeare's Wordplay* (London: Methuen, 1957) 104–105.

[59] Flora Haines Loughead, *Dictionary of Given Names* (Glendale, CA: Arthur Clark Co., 1934) 137, 309, 78.

[60] Patrick Grant, "The Magic of Charity; A Background to Prospero," *Review of English Studies* 27 (1976): 1–16.

[61] *The Oxford English Dictionary* (Oxford: Clarendon, 1933) Vol. XI, 369.

[62] *OED* vol. XI, 369.

[63] *OED* vol. X, 934.

[64] *OED* vol. II, 31.

[65] Alden T. Vaughan, "Shakespeare's Indian: The Americanization of Caliban," *Shakespeare Quarterly* 39 (1988): 138.

[66] Loughead 34.

[67] *OED* vol. IX, 346.

[68] Loughead 29.

[69] *OED* vol. I, 359–60.

[70] *OED* vol. I, 377.

[71] Guiseppe Ragazzini, ed., *The Follett / Manichelli Italian Dictionary* (Chicago: Follett, 1967) 1242.

[72] David S. Berkeley, Lecture, Shakespeare Class, Oklahoma State University, Stillwater, Oklahoma, 31 Oct. 1977.

[73] *OED* vol. XI, 165.

[74] Law 3.

[75] Law 3.

[76] Law 7.

[77] Unwin 179.

[78] Wright 10.

[79] Kraus 39.

[80] Thomas Ingelend, *The Disobedient Child*, ed. John S. Farmer, *The Dramatic Writings of Richard Wever and Thomas Ingelend* (New York: Barnes and Noble, 1966) 43–92.

[81] Dekker, ed. Bowers, 282.

[82] Kraus 40.

[83] Grant 1–16.

[84] Gerald Eades Bentley, "Shakespeare and the Blackfriars Theater," *Shakespeare Survey* 1 (1948): ed. Allardyce Nicoll (Cambridge: UP, 1948): 38–49.

[85] William Armstrong, "Acting and Theaters," *Shakespeare Survey* 17 (1964): 191–204.

Claudius, the Villein King of Denmark

by David S. Berkeley

In the Shakespearean plays are, genetically speaking, only two classes of men, gentles and base, the latter not possessing coats of arms.[1] Criticism, clamped in the vise of democratic obliviousness to blood-based class distinctions, has not noticed that Claudius, by virtue of nine times being named a "villain" or "villein," is thus intimated to be base, and therefore, being a member of the royal family, a bastard half-brother to Hamlet's father. The idea of a villein king of Denmark, profoundly disturbing to Hamlet (and presumably to Shakespeare), is the most notorious violation of degree in the realm and must have frightful consequences until a legitimate king, qualified by having the purest blood in Denmark and by "kindly enclyning" all those beneficent qualities that attend it, is seated on the nation's throne. The idea of Queen Gertrude's being married to a villein is perhaps even more disruptive of Hamlet's equanimity because it means that night after night she is becoming less consanguineous with her son, in effect less his mother: her physiological system is gradually losing the thin, hot, fast-flowing, sweet-tasting red blood of Hamlet's father (and as a result all his excellences, physical and moral) and is becoming villeinized with Claudius' blood (and therefore acquiring all his vicious and ugly qualities). Failure to grasp this conception renders the play more mysterious than it really is by not fully accounting for Hamlet's anger.

Shakespeare's weed-imagery—weeds being a metaphor for the base—implicates Claudius as a villein. In his first soliloquy Hamlet figures his *Weltschmerz* thus: "'Tis an unweeded garden, / That grows to seed, things rank and gross in nature / Possess it merely."[2] "Rank" and "gross" are words that Shakespeare elsewhere associates with the baseborn: "weeds" is a metonymy for the base just as "lilies" is its antonym for gentles in "Lilies that fester smell far worse than weeds" in Sonnet 94. So in the closet scene Hamlet, believing his mother to be a degenerating gentle who is open to repentance, urges her not to "spread the compost on the weeds / To make them ranker." The "rank" and "gross" baseborn could not in Shakespeare's plays be lifted to a higher blood level (except in Henry V's French expeditionary force); so Hamlet at no time attempts to talk Claudius out of his evil courses.

Claudius' unprepossessing appearance suggests his peasant origins. Hamlet analogizes his father to "Hyperion," his uncle to a "satyr." The Ghost of his father compares Claudius to "garbage," itself to "a radiant angel." In the closet scene Hamlet bids his mother see in her late husband "Hyperion" again and likens her present husband to a "mildewed ear" and a "moor." One might suppose that Hamlet compromises the gravity of his charges of murder and incest against Claudius by paralleling them to an aesthetic (and impressionistic) evaluation of Claudius' appearance. Not so. Physical appearance has in the Shakespearean plays and poems an ethical dimension. Tarquin, for example, apprehends that his rape of Lucrece is "so vile, so base, / That it will live engraven in my face." Personal unrighteousness of parents stamped their offspring with ugly, peasant-like features. Thus Prospero warns Ferdinand and Miranda that premarital sex will cause their children to resemble "weeds," i.e. villeins (*The Tempest* IV.i.19–22): "barren hate, / Sour-eyed disdain, and discord shall bestrew / The union of your bed with weeds so loathly / That you shall hate it both." Of Claudius we gather no details except that he is dark, being likened to a "satyr" and a "moor": a coloring not esteemed in the Shakespearean plays. Claudius, we gather, has the superficies of an animal-like churl. Since "beauty breedeth beauty" (*Venus and Adonis,* line 167) and since no one in the play intimates that Claudius derogated from noble parents, one gathers that he is congenitally base.

Claudius' lack of honor suggests his baseness. He calmly stomachs Laertes' insulting him in public with the imputation of villein birth—"O thou vile King." When his conniving with Laertes at Hamlet's death misfires, he vilely submits, in spite of the presence of friends and guards, to Hamlet's degrading coercion. He specializes in villainy, in low animal cunning, in personally spilling defenseless blood, or in bidding others serve his bloody purposes.

Claudius, we gather, smells. In attempting to dissuade his mother from further intercourse with Claudius, Hamlet depicts this act as taking place "In the rank sweat of an unseamed bed, / Stewed in corruption, honeying and making love / Over the nasty sty." The "rank sweat" must come primarily from Claudius, and it implicates him in baseness as it does Cloten in *Cymbeline* II.i.17. This "demo stench,"[3] to use Brents Stirling's word, imports "villainous melancholy" (*King Lear,* I.ii.147), the ineradicable humoral smell of the base[4]—something never alleged of the gentle in Shakespearean plays. Claudius, although dieting on high fare for a lifetime, has not been able to alter his humoral smell for the better.

A more important evidence for Claudius' baseness is Hamlet's nine times branding his uncle with this most scalding of epithets in Shakespeare's cursing vocabulary—"villain." After meeting the Ghost of his father, Hamlet says of Claudius, "O villain, villain, smiling, damned villain ... That one may smile, and smile, and be a villain." Villains, sometimes called "honest," were thought to verbalize anything that passed through their consciousness. Hamlet here means that even a vile person pursuing his villeinous courses—Claudius—can mask them by smiling—behavior rare enough to deserve memorializing in a daybook. Hamlet later brands his uncle, after a markedly class-conscious description, "A murderer and a villain, / A slave that is not twentieth part the tithe / Of your precedent lord, a vice of kings" (III.iv.96–98). If one takes *villain* as "a very wicked man," as critics of *Hamlet* do, then Hamlet is saying, jejunely, that Claudius is "a murderer and a very wicked man." But Hamlet is really saying "a murderer and a bastard" because "villain," *OED,* 1 ("a low-born base-minded rustic"), and "villein," *OED,* 1 ("one of the class of serfs in the feudal system"), then interchangeable words and identical in pronunciation, both denoted base, bastard birth. C. S. Lewis has pointed out that "villain" in *Hamlet* "still carries some implication of ignoble birth ...." *(Studies in Words)*: "We call the enemy not what we think he is but what we think he would least like to be called."[5] What Lewis does not notice is that Hamlet never *vis-a-vis* brands Claudius "villain." The insult loses all potency in soliloquies, where it almost always appears. To be stigmatized a villain, class sense, is for all but a few men like Henry VI to endure a scorn far more scathing and more shameful than to be called a villain, ethical sense. To be a villain, class sense, i.e., a villein, is to be a base bastard without bar sinister because such a person has a base father, therefore no coat of arms (unlike the Bastard in *King John*). To be a villain, class sense, places horns on the putative father's forehead, impeaches the chastity of the mother, and intimates that the villain's brothers and sisters may be only half-brothers and half-sisters, and on this account little love and no honor may be found in the family. To be called a villain, ethical sense, is not in the highest degree vituperative because the person so described may impute his villainy to necessity, to some external pressure, as evil men have traditionally done. To brand Claudius or any other king a villain, class sense, is to impugn his legitimacy and in effect to demand his abdication. Gertrude, whom Hamlet accuses of knowingly marrying a villein, hears these class-insults without reproach or dispute. (Those doubting this argument have the impossible task of demonstrating that the words "villain" and "villein" had no class meaning circa 1600.) Laertes plasters the class-adjective "vile" on Claudius' forehead—"O thou vile King"—a word almost certainly regarded in relation to "villain" as "base" was to "bastard." The King, lacking gentlemanly choler, makes no remonstrance at this insult direct, nor does the Queen. The King is never called "gentle" although even actors are called "Gentlemen" (II.ii.387). His role in the dumbshow is taken by a "fellow," Shakespeare's class-word for the murderer.

When Shakespeare wishes to establish a child's legitimacy, he invokes the Renaissance maxim of *simili generari a simile,* causing remark of children's printing off their fathers, of which the most egregious example is Banquo's hirsute line. Other examples of children bearing close physical resemblance to their sires and on this account being regarded as legitimate are found in the Duchess of York's protestation regarding Aumerle in *Richard II* V.ii.79 and in Paulina's berating of Leontes concerning the infant Perdita in *The Winter's Tale* II.iii.97–103. Obviously Claudius and Hamlet *pere* could not both resemble their father. If we believe Hamlet—we have no other choice—Hamlet *pere* must resemble his father, and Claudius, who differs *toto caelo* from the elder Hamlet, must resemble someone else. Claudius' ruling passion is envy: as Bacon wrote, "... bastards are envious, for he that cannot possibly mend his own case will do what he can to impair another's" ("Of Envy").

The implications of Claudius' bastardy are portentous. The gravamen lies upon Hamlet's paternal grandmother, unmentioned in the play because of the extreme delicacy of Hamlet's and his father's positions. Hamlet generalizes his grandmother's weakness in arraignments of the female sex. The antecedent action appears to be this: Hamlet's grandmother after giving birth to Hamlet's father, a true son of Hamlet's grandfather on the spear side, formed an adulterous alliance with a man of the villein class, and their child was Claudius, half-brother to Hamlet *pere.* The father, not the mother, determined class-status as with Shakespeare himself. So a villain in the class sense is seated on Denmark's throne in true-born Hamlet's place.

Another implication, possibly worse from Hamlet's point of view, is that Gertrude is becoming villeinized as through intercourse with her new husband she acquires Claudius' blood and loses the blood of Hamlet's father. Period medical thought is summarized by Jacques Ferrand in *Erotomania; or a Treatise of Love*: "The seed is nothing else but blood made white by the naturall heat."[6] Claudius and Gertrude are "one flesh," therefore one blood. This conception, I believe, is unknown to criticism of *Hamlet.*

The consequences of Claudius' bastardy need assessing. Inasmuch as he is only a half-brother-in-law to Gertrude, Shakespeare mutes the theme of incest and magnifies Gertrude's tasteless susceptibility to Claudius' sexual overtures. Hamlet's tirade against his mother derives some of its fury from his fear that she is at or near the point of no return to gentility: one does not lecture peasants for being possessed of bad blood. Hamlet, like Polixenes in *The Winter's Tale,*[7] has his reasons: who knows what iniquity, what dishonor, what ineptitude, and what foolishness lie in the base dunghill sanguinary pedigree of Claudius. So Hamlet insults Ophelia, a potential dissolver of gentle families. Another consequence is Claudius' poltroonish behavior during the rapier scene. Hamlet, though fatally poisoned by Laertes' rapier, has by the strength of his superior blood, to some degree resistant to toxins,[8] run Laertes' weapon not only through Laertes (who having less pure blood dies first) but also through the unwounded, unfatigued, and guarded Claudius. Hamlet forces the King to drink off the poisoned wine as if he were medicining a dog, an act that is finely symbolic not only of Claudius' animal-like appearance and subhuman inner qualities but also of wine as blood: wine, virtually sanguified in period understanding,[9] here symbolizes that Claudius is obliged to drink his own poisonous blood now that it has killed the Queen. The King could have drawn his own rapier to protect himself, he could have commanded his guards to kill Hamlet, he could have at least clamped his jaws against the poison; but, no gentleman, he knows not honor.

The ending of *Hamlet* is another exhibition of natural hierarchy. Shakespearean plays teach that all is well with the world when the best blood is enthroned, when gentles are properly placed, when the base are carrying on their often-mechanically oriented occupations; the world is out of joint when these happy conditions do not prevail.

Notes

1 For a fuller account of Renaissance understanding of genetics and physiology in relation to Shakespeare's characterization, see David Shelley Berkeley, *Blood Will Tell in Shakespeare's Plays*, Graduate Studies Texas Tech University 28 (Lubbock: Texas Tech UP, 1984).

2 Shakespearean quotations in this paper usually follow the text and lineation of *Shakespeare: Major Plays and the Sonnets*, ed. G. B. Harrison (New York: Columbia UP, 1949).

3 *The Populace in Shakespeare* (New York: Columbia UP, 1949).

4 Lanfranc and Robert Burton mention this humoral smell. "Þer is engendrid anoþer substance þat is sumwhat stynking, & is clepid malancholi," states Lanfranc's *Ciruge*, 202, ca. 1400 (*OED* 1 b). Burton notes that "fumes ... arise from this corrupt blood [to] disturb the mind," melancholiacs being "much troubled with wind" and proneness to "continual, sharp, and stinking belchings" (*Anatomy*, Part i, Sect. 3).

5 (Cambridge: Cambridge UP, 1960).

6 (Oxford, 1649) 261. Thus Eucharius Roesslin, *The Byrthe of Mankynde*, trans. R. Jonas (London, 1598), Capit. IX: " ... the sparme or sede of man or woman being attracted by the foresaid seed bringers into the stones, passeth by many narow strayghtes, which being before vayne bloud and artere bloud, by vertue of the places through which it had to passe, becommeth from red colour to whyte."

7 David S. Berkeley and Zahra Karimipour, "Blood-Consciousness as a Theme in *The Winter's Tale,*" *Explorations in Renaissance Culture* 11 (1985): 89–98.

8 Cf. John Trapp, *A Commentary or Exposition Upon all the Books of the New Testament* (1647), ed. W. Webster (1865; rpt. Grand Rapids: Baker Book House, 1981) 487: "No more ["and felt no harm"] did Queen Elizabeth, when Squier, the traitor, sent by Walpool the Jesuit, had poisoned the pummel of the saddle. The vigor of the poison (said the Jesuit) is such, as neither continuance of times nor subtilty of air could check or unvirtuate. And yet, albeit the season were hot, and the veins open to receive any malign tainture, her boy thereupon felt no distemperature, nor her hand felt no more hurt (saith Speed) than Paul's did when he shook off the viper into the fire."

9 Tobias Whitaker, *The Tree of Humane Life, or the Bloud of the Grape* (London, 1638) 30: "wine, especially Claret or red, is sanguified before it is taken."

Publications of David S. Berkeley

Shakespeare

"Antony, Cleopatra, and Proculeius." *Notes and Queries* 195.25 (December 9, 1950): 534–35.

"The Crux of *Antony and Cleopatra*," Bulletin of the Oklahoma Agricultural and Mechanical College, Arts and Sciences Studies, Humanities Series 4 (1959).

"On Oversimplifying Antony." *College English* 17.2 (1955): 96–99.

Defense of thesis of the paper above in a letter to the editor of *College English* 18.5 (1957): 286–87.

"A Vulgarization of Desdemona." *Studies in English Literature* 3.2 (1963): 233–39.

"'Determined' in *Richard III*, I.i. 30." *Shakespeare Quarterly* 14.4 (1963): 483–84.

"On Desentimentalizing Antony." *Notes and Queries* N.S. 11.4 (1964): 138–42.

With Donald Eidson. "The Theme of *Henry IV, Part I*." *Shakespeare Quarterly* 19.1 (1968): 25–31.

Blood Will Tell in Shakespeare's Plays. Lubbock, TX: Texas Tech Univ. Press, 1984.

With Zahra Karimipour. "Blood-Consciousness as a Theme in *The Winter's Tale*." *Explorations in Renaissance Culture* 11 (1985): 89–98.

"Claudius the Villein King of Denmark." *Hamlet Studies* 11.1 (1989): 9–21.

With Donald Keesee. "Bertram's Blood-Consciousness in *All's Well That Ends Well.*" *Studies in English Literature* 31 (1991): 247–58.

With Donald Keesee. "Bertram's Blood-Consciousness in *All's Well That Ends Well.*" Reprinted in *Shakespearian Criticism* 38 (1977): 155–60.

Milton

"*Precieuse* Gallantry and the Seduction of Eve." *Notes and Queries* 196.16 (1951): 337–39.

"'Determinate Sentence' in Milton's *Of Education.*" *Notes and Queries* N.S. 1.1 (1954): 25–26.

"An Unnoted Biblical Echo in Milton's Sonnet on the Massacre in Piemont." *The Explicator* June, 1957: Item 58.

"The Revision of the Orpheus Passage in *Lycidas.*" *Notes and Queries* N.S. 5.8 (1958): 178.

"A Possible Biblical Allusion in *Lycidas,* 1." *Notes and Queries* N.S. 8.5 (1961): 178.

"Thematic Implications of Milton's Paradise of Fools." *Papers on Milton.* The University of Tulsa Department of English Monograph Series 8 (1969): 3–8.

Inwrought With Figures Dim: A Reading of Milton's "Lycidas." The Hague: Mouton, 1974.

"Dalila as Amphisbaena." *Papers on Language and Literature* 19.1 (1983): 87–92.

"On a Common Error Respecting Samson's Size and Musculature." *English Language Notes* 19.1 (1983): 260–262.

With Salwa Khoddam. "Samson the Base Versus Harapha the Gentle." *Milton Quarterly* 17.1 (1983): 1–7.

"'Light' in Milton's Sonnet XX." *Philological Quarterly* 61.2 (1982): 208–211.

With Salwa Khoddam. Rejoinder to L. Miller on *Samson Agonistes* (*Milton Quarterly* 18 [1984]: 25–27). *Milton Quarterly* 18.4 (1984): 129–31.

With Salwa Khoddam. Rejoinder to N. Flinker. "Biblical Samson: More About the Hero" (*Milton Quarterly*, 19.1 [1985]: 19–21). *Milton Quarterly* 20.2 (1986): 54.

"The 'Mysterious' Marriage of Adam and Eve in *Paradise Lost.*" *Philological Quarterly* 66 .2 (1987): 195–205.

"Michael's New Commandment 'With Promise': *Paradise Lost* XI, 530–546." *Papers on Language and Literature*, 24.2 (1988): 134–41.

With Byung-Eun Lee. "The Evil 5: Numerological Symbolism in *Paradise Lost.*" *English Language Notes* 27.4 (1990): 33–38.

Reply to Dennis Daniel, "Through the Telescope of Typology." *Milton Quarterly* 26 (1992).

Restoration Drama

"An Early Use of 'Sensibility.'" *Notes and Queries* 193 (1948):150.

"The Penitent Rake in Restoration Comedy." *Modern Philology* 49 (1952): 223–33.

"*Préciosité* and the Restoration Comedy of Manners." *Huntington Library Quarterly* 18 (1955): 109–28.

"Some Notes on Probability in Restoration Comedy." *Notes and Queries* N.S. 2 (1955): 342–44.

"The Art of 'Whining' Love." *Studies in Philology* 52 (1955): 478–96.

"Sir Walter Scott and *Préciosité.*" *Nineteenth Century Fiction* 10 (1955): 240–42.

"The *Précieuse,* or Distressed Heroine, of Restoration Comedy." Bulletin of the Oklahoma State University, Arts and Sciences Studies; Humanities Series 6 (1959).

Linguistics

"An Unnoted Sense of 'Whining.'" *American Speech* 27 (1952): 225–28.

"An Acceptable Comma Splice." *College English* 14 (1953): 289.

"Agreement of Subject and Verb in Anticipatory *There* Clauses." *American Speech* 29 (1954): 92–96.

"A Euphemistic Sense of 'Exotic' in Oklahoma." *American Speech* 30 (1955): 79–80.

"A Current Undefined Sense of *Impress, Unimpressed,* and *Impression.*" *American Speech* 29 (1954): 192–95.

"An Unclassified Past Tense of *Fit.*" *American Speech* 30 (1955): 311.

"Agreement of Subject and Verb in Anticipatory *There* Clauses." *Readings in Applied English Linguistics,* ed. Harold B. Allen. (New York: Appleton-Century-Crofts, 1958): 250–54.

Typology

"*Figurae Futurarum* in *Moby-Dick.*" *Bucknell Review* 21 (1973): 108–23.

"Some Misapprehensions of Christian Typology in Recent Literary Scholarship." *Studies in English Literature* 18 (1978): 3–12.

Article on "Typology." *Dictionary of Biblical Traditions in English Literature,* ed. David L. Jordan (Grand Rapids: Eerdmanns, 1992): 792.

Miscellaneous Writings in Prose

"On the Indiscriminate Use of Quotations." *College English* (1955): 370–71.

"On the Definition of 'Tragic Irony.'" *Notes and Queries* N.S. 1 (1954): 521–22.

"A Vulgar Error Touching Calvin's Doctrine of Total Depravity." *Notes and Queries* 196 (1951): 293–95.

Article on "Allegory." *Dictionary of the Biblical Tradition in English Literature,* ed. David L. Jordan (Grand Rapids: Eerdmaus, 1992): 29–31.

"Is a Classical Education Obsolete? An Elitist Response." *Cimarron Review* 79 (1987): 5–12.

Miscellaneous Writings in Verse

"A Brighter Hellas." *Christianity Today* 3 (April 13, 1950).

"Io Triumphe." *National Poetry Association Anthology* (Los Angeles, 1960).

"Science at the Second Coming." *Christianity Today* 15 (May 1971).

"In Memoriam Caroli Obadiah Beery." *The Strand Book of Poetry 72* (London: Florwain Publishers, 1972).

"The Oklahoma 4-H Convention." *The Strand International Book of Modern Poets* (London: Florwain Publishers, 1973).

"Churchill." *Yearbook of Modern Poetry 1976* (Knoxville: Young Publications, 1976).

"On C.S. Lewis." *Bulletin of the New York C.S. Lewis Society* 14. 3 (1983): 8.

"On Robbie Hiner's Singing." *Silver Wings.* Date unknown.

Notes on Contributors

Dennis F. Bormann earned the PhD in English from Oklahoma State University in 1994. A published author of fiction and poetry, and the possessor of the MFA degree from Vermont College, Mr. Bormann resides in Stillwater, Oklahoma, with his wife and children.

John W. Crawford has been Professor Emeritus of English, Henderson State University in Arkadelphia, Arkansas, since 1997. His books include *Romantic Criticism of Shakespearian Drama,* University of Salzburg Press, 1978; *Discourse: Essays on English and American Literature,* Rodopi Press, 1978; *Early Shakespearean Actresses,* Peter Lang, 1984; *The Learning, Wit, and Wisdom of Shakespeare's Renaissance Women,* Edwin Mellen Press, 1997.

His articles on Renaissance literature have been published in *SCMLA South Central Bulletin, The Southern Quarterly, University of Dayton Review, The Upstart Crow, Explicator, Philological Association of Mississippi Publication, North Carolina Journal of Speech and Drama,* and *Philological Association of Arkansas Publication,* among others. His books of original poetry include *Making the Connection,* Edwin Mellen Press, 1989; *I Have Become Familiar With The Rain,* Edwin Mellen Press, 1995; *Winning Poems* (Chapbook), Poetry Forum Press, 1992.

Esther M. Gloe, now retired, was Assistant Professor of English at the University of Southern Colorado, where she taught courses in technical writing, Shakespeare, Western world literature, and composition. She received her PhD from OSU in 1979; her dissertation is *The Influence of the English Work Laws on the Drama of the Period from 1563 to 1642.* Her article, "Setting Up Internships in Technical Writing," appeared in the *Journal of Technical Writing and Communication* 13.1(1983). Her presentations at the conventions of the Popular

Culture Association include these: "Machines Used as Symbols in Cozzens' Novel, *S.S. San Pedro,*" "Ethnic Clashes Reflected in Our Popular Entertainment," and others. Since retirement, she has worked as a technical writer, her articles have appeared in newspapers and magazines, and her poems have appeared in literary journals.

Kurt Hochenauer is an Associate Professor of English at the University of Central Oklahoma, Edmond, where he teaches courses in modern British literature and serves as Director of English Graduate Studies. His articles have appeared in such journals as *Studies in the Novel* and *The Midwest Quarterly.* His short fiction and poetry have appeared in several journals, including *The Missouri Review* and *Best Minds.* Most recently, he has completed a collection of poetry.

Myung-Soo Hur was graduated from Kyung Hee University with the BA (1976) and MA (1978) degrees in English literature. He studied at Oklahoma State University from 1986 to 1993, receiving the MA in 1989 and the PhD in 1993. He teaches English at Kyung Hee University in Seoul, Korea.

Byung-Eun Lee was awarded the PhD in English from Oklahoma State Univeristy in 1994 with his dissertation "Milton's Class Consciousness." Along with Professor Berkeley, he has published "The Evil Number Five: Numerological Symbolism in Paradise Lost" in *English Language Notes* 27.4 (1990): 33–38. Mr. Lee presently serves as Assistant Professor of English Language and Literature at Hansung University in Seoul, Korea.

Dilin Liu is an associate professor of linguistics/TESL at Oklahoma City University where he teaches English, linguistics, and TESL. He has published on American and English literature, linguistics, and English education. His articles have appeared in journals such as *English Today: The International Review of the English Language, The ELT Journal: An International Journal for Teachers of ESOL, International Journal of Intercultural Communication, TESOL Quarterly,* and *Short Story.* He also has a few articles in edited anthologies. Currently, he is working on several articles on applied linguistics and a book entitled *Metaphorical Language and Construction of Life and Reality: The Case of American English and the Chinese Language.*

Shirley Marney is Senior Instructor at Texas State Technical College, where she specializes in composition and technical communications. She regrets that her limited time offers little opportunity for attention to the sort of intellectual stimulation found in literature. What she describes as her "prosaic existence," however, has only intensified her lifelong love of Shakespeare and her conviction that he speaks even to a generation dedicated to technology.

Randy Phillis is an Associate Professor of English at Mesa State College, and Chair of the Department of Languages, Literature and Communications. His poetry has appeared widely in journals, including *The Iowa Review, South Carolina Review, Passages North,* and *Denver Quarterly*. His collection, *A Man Explains His Posture,* was published in 1994. He edits *Pinyon Poetry,* The Frank Cat Press, and is poetry editor for *National Forum.* He recently completed editing an anthology of "Generation X" writers which is scheduled for publication in 1999.

Peter C. Rollins is Regents Professor at Oklahoma State University, where he served as a colleague of Dr. Berkeley. Other books edited by Rollins include *Hollywood as Historian* (2nd ed. UP of Kentucky, 1998); *Hollywood's World War I* (Popular Press, 1998); *Hollywood's Indian* (UP of Kentucky, 1999). He edits *Film & History: An Interdisciplinary Journal of Film & Television Studies,* a publication which has a web site at http://h-net.msu.edu/~filmhis/. As this book goes to press, he is bringing to completion a major reference work titled *The Columbia Companion to American History on Film* (Columbia UP, 1999).

Alan Smith is Director of English Studies at City College of Norman, Oklahoma. His articles on Marlowe, Shakespeare, Milton, and Conan Doyle have appeared in such journals as *American Notes and Queries, Explicator,* and *Victorian Newsletter.* During recent years he has devoted much of his time to painting and has participated in solo and group art shows in Norman and Oklahoma City.

Kenneth J. Tiller is assistant professor of English at Clinch Valley College of the University of Virginia, where he teaches Chaucer, the Medieval romance, and Arthurian literature. His main area of interest is Early Middle English literature, especially Law$_3$amon's Brut. He has also written on Malory and on the *Anglo-Saxon Chronicles.*

Index

STUDIES IN SHAKESPEARE

Edited by Robert F. Willson, Jr.

This series deals with all aspects of Shakespearean drama and poetry. Studies of dramatic structure, verse and prose style, major themes, stage or performance history, and film treatments are welcomed. The editor is particularly interested in manuscripts that examine Shakespeare's work in its American setting—in the academy, on stage, and in popular culture. Inquiries and manuscripts should be sent to the series editor:

Robert F. Willson, Jr.
Department of English
University of Missouri-Kansas City
College of Arts & Sciences
106 Cockefair Hall
Kansas City, MO 64110-2499

To order other books in this series, please contact our Customer Service Department at:

(800) 770-LANG (within the U.S.)
(212) 647-7706 (outside the U.S.)
(212) 647-7707 FAX

or browse online by series at:

WWW.PETERLANG.COM